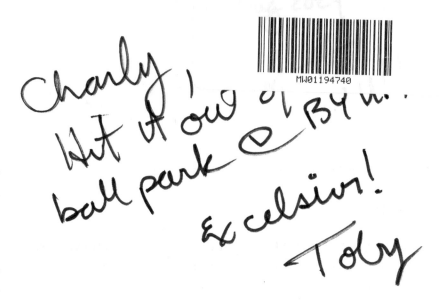

Connecting the Dots
Building Your Network and Legacy

Toby Usnik and Samir Kanuga

Blue Ocean Press

Copyright @ 2024 Blue Ocean Press

All Rights Reserved.

This publication may not be reproduced, stored in a retrieval system or transmitted in any form or by any means, electronic, mechanical, photocopying, recording, or otherwise, without prior written permission of the publisher, except by a reviewer who may quote brief passages in a review to be printed in a periodical.

Published by:
Blue Ocean Press

U.S. Office
P.O. Box 510818
Punta Gorda, Florida 33950

URL: http://www.blueoceanpublications.com
Email: books@blueoceanpublications.com

ISBN: 978-4-902837-64-3

Table of Contents

Introduction: Connecting the Dots	5
Chapter 1: The Importance of Connections	23
Chapter 2: Intergenerational Views on Networking	33
Chapter 3: Mixing: Connecting with Anyone	65
Chapter 4: Matching: Connection More Strategically	87
Chapter 5: Diversity	127
Chapter 6: Trust: The Gold Standard	141
Chapter 7: Barometers of Trust	165
Chapter 8: Cross-Check: Nurturing Your Relationships As You Navigate Life	183
Chapter 9: Networking in the Digital Age	199
Chapter 10: Tools in Your Toolbox	215
Chapter 11: Old School Matters	245
Chapter 12: The Power of Connection: Optimizing Your Network and Legacy	265
Acknowledgments	287
About the Authors	289

Introduction

The expression "connecting the dots" can be about discerning patterns in nature or finding ways to move. Just as we can connect dots in our actions, we can connect with people in our lives, less accidentally and more deliberately, allowing us to discover our own purpose in life and create our most amazing legacy. Who we are associated with in life speaks volumes about who we are, what we stand for, and why our lives matter. Those associations will also represent us after we die, essentially defining our legacies. As such, we think of networking as the professional term for the skill of relationship building, a skill that we should all learn to enhance society.

Defining who we are, who we connect with, how we connect, and what we do with those connections are the aims of this book.

This book is a guide and inspiration to living your best life while catalyzing the same in others, one relationship at a time. It requires an open mind and a desire to exercise your interpersonal skills. To that end, we even experimented with AI in creating this book to exemplify that the principles of human connection are timeless while technologies are simply tools to help or hinder connection. You, dear reader, are the judge of how well we have done. You, too, are the judge of how best to connect the dots of your network and build your legacy.

The way we connect with others has undergone a remarkable transformation in recent years, driven by the relentless advancement of technology. From the cumbersome Rolodexes of the past to the sophisticated AI-powered tools of today, networking has evolved into a dynamic and ever-changing landscape.

Gone are the days of meticulously maintaining physical address books and relying on word-of-mouth referrals to expand one's personal or professional circle. Today, a vast array of digital tools and platforms has emerged, offering unparalleled opportunities to connect with individuals across industries, geographical boundaries, demographics, and even time zones.

The rise of social media giants like LinkedIn and Facebook has revolutionized professional networking, providing a centralized platform for individuals to showcase their expertise, engage with potential collaborators, and discover new opportunities. Mobile applications have further streamlined the process, allowing users to manage their networks and make connections on the go.

But the most significant technological leap in networking lies in the realm of artificial intelligence (AI) and algorithms. AI-powered tools are now capable of analyzing vast amounts of data to identify potential connections, recommend relevant networking events, and even provide personalized advice on building and maintaining relationships.

This technological revolution continues to transform networking from a mere formality into a powerful tool for achieving personal and professional goals. Individuals can now leverage AI to expand their reach, uncover hidden opportunities, create in unprecedented ways, and cultivate meaningful connections that can propel their lives and careers forward.

Congratulations on taking greater ownership of your network and legacy by connecting the dots with us.

Introduction: Rationale for Connecting the Dots

With this book, we have collected 100 years of lived experience on human interaction. We have combined this with extensive research on human networking and interviews with notable connectors to distill for our readers the best practices for growing your personal and professional networks to help optimize the life that you lead, and the legacy that you will leave behind.

Chapter 1: The Importance of Connections

This chapter discusses the importance of connection in our lives. We explore the benefits of having strong relationships, both personal and professional. We explore leading purposeful lives and generating a positive wake as your legacy. Learn from famous connectors, young and old.

Chapter 2: Intergenerational Views on Networking

This chapter explores the different views on networking that different generations apply in their lives. We discuss the challenges and opportunities that each generation faces when it comes to networking, and we provide advice on how to network effectively across generations. Learn from such famous connectors as Gen Zers Billie Eilish, Greta Thunberg, and Emma González; millennials Mark Zuckerberg, Jeff Bezos, and Sundar Pichai; Gen Xers Barack Obama, Sheryl

Sandberg, and Elon Musk; and baby boomers Oprah Winfrey, Bill Gates, and Warren Buffett.

Chapter 3: Mixing: Connecting with Anyone

This chapter provides practical advice on how to connect with anyone. The universe sends all kinds of people into our lives each day, so how best to engage with them? We cover topics such as how to start a conversation, how to build rapport, and how to stay in touch.

Chapter 4: Matching: Connecting More Strategically

This chapter delves into the art of strategic networking, empowering you to navigate the intricate dance of building relationships that advance your personal and professional goals. By understanding yourself and your purpose in life, you will be more effective in adventuring out into the world, one relationship at a time. In an increasingly interconnected world, the ability to forge meaningful connections is more valuable than ever. Discover how to identify and target individuals who align with your aspirations, craft compelling introductions that capture attention, and engage in conversations that foster mutual understanding and trust. Learn techniques to amplify your social presence, leverage technology to expand your reach, and transform casual encounters into lasting partnerships.

Chapter 5: Diversity

Diversity is important in every aspect of our lives, and networking is no exception. When we connect with people from different backgrounds, we expose ourselves to new ideas and opportunities. This can help us grow as individuals and achieve our personal and professional ambitions. This chapter summarizes the benefits of having a diverse network in the world.

Chapter 6: Trust: The Gold Standard

This chapter discusses the importance of trust in relationships, digital and physical. It provides tips on how to build trust, and it explains how to repair trust when it has been broken. You build a reputation over time, but it can be destroyed in a moment. Learn the concept of reputational currency and use it as you build your network and legacy.

Chapter 7: Barometers of Trust

This chapter explores our digital age when digital connections permeate our lives and the search for purpose and spiritual meaning remains constant. Navigating spaces like dating apps, horoscopes, and faith communities can feel like a trust-tightrope walk.

Chapter 8: Cross-Check: Nurturing Your Relationships

This chapter discusses the importance of maintaining relationships. We provide tips on how to stay in touch with friends, family, and colleagues, and we explore how to nurture relationships over time. We also share why it is okay to let some relationships go.

Chapter 9: Networking in a Digital Age

In this chapter, we pull back the curtain to reveal the intricate workings of algorithms, the unseen forces driving social media recommendations. Delve into the fascinating world of artificial intelligence (AI) and its role in shaping your online connections. Discover how algorithms analyze your behavior and interactions to identify potential connections that align with your interests and goals. Uncover the hidden biases and limitations of these algorithms and learn how to navigate the digital landscape with awareness and discernment. Empower yourself to harness the power of algorithms and train them to expand your network, uncover hidden opportunities, and cultivate meaningful connections in the ever-evolving digital age. You've got the power.

Chapter 10: Tools in Your Toolbox

This chapter discusses how to network effectively in the digital age. We cover topics such as how to use AI in everything, connect with others via social media, attend online events, and network virtually. Learn from such famous connective platforms as LinkedIn, Twitter, and Facebook.

Chapter 11: Old School Matters

This chapter discusses how to glean the best of old-fashioned networking practices while not abandoning the digital world entirely. We help you find a healthy balance. Use the power of the internet to reach out and connect, and then nurture those connections through the warmth of personal interaction. You will see how such notable connectors as Kevin Bacon and David Rockefeller became such successful networkers. We invite you to still go ahead, pick up the phone, schedule that coffee date, and watch your career flourish under the sunshine of genuine connection.

Chapter 12: The Power of Connection: Optimizing Your Network and Your Legacy

This chapter discusses the power of connection in our lives. We explore how connection can help us live happier and more successful lives. We aim to help you connect the dots in your network for a more purpose-driven life and legacy. Following the philosophy, tips, and life hacks outlined in this book, you will lead a more fulfilling life and be remembered how you touched others.

Introduction: Connecting the Dots

Who are you? Who are your people? What do you do? What is your purpose? What will you be remembered for?

We all address these ageless and important questions in life, usually reactively rather than proactively. Most people tend to let happenstance determine their answers. In *Connecting the Dots*, we will show you methods and examples of ways to answer these major questions proactively, honestly, efficiently, and happily. We will also show you how to marry your findings with human networks to achieve a life well lived.

We start by exploring the network of people that you know, or want to know, and seeing the pitfalls and opportunities that your deliberate exploration can reveal.

In today's rapidly evolving world, networking and interpersonal engagement are not just valuable skills, but essential tools for success and happiness. In a world increasingly driven by connection and collaboration, the ability to effectively build and maintain relationships is crucial for both personal and professional growth.

Strong networks provide access to a wealth of knowledge, resources, and opportunities. They open doors to new ideas, collaborations, and partnerships that can fuel growth. Moreover, in a world where technology can sometimes isolate us, meaningful interpersonal engagement fosters a sense of belonging and community, which is essential for our resilience.

Networking, sometimes called interpersonal engagement, is not just about building social capital; it is also about respect for others. By connecting with people from diverse perspectives, one creates their own increasingly interconnected world.

For us, networking is also about quality over quantity. Having huge followings on social media or trending on TikTok may give an illusion of being relevant or important to others, but it is not the same as being the trusted, go-to person that others rely on for advice or support.

As you read this book, ask yourself to what end you connect with people day in and day out. If the answer is not evident, dig deeper. Are you simply focused on a transactional relationship, such as being an employee earning a wage or salary? Are you connecting for fun and socializing and if so, why do you do so with those people you choose? By asking yourself why you connect with someone will shed light on how open minded you are to different kinds of people. By asking yourself why you stay connected with someone will be equally illuminating because it represents the most basic starting point to building meaningful and lasting relationships.

In a time of uncertainty and change, the ability to connect with others is more important than ever. It also requires greater effort to sustain those relationships despite the illusion of connectivity that social media has given us. By deliberately investing in our future, we build the foundation for a more collaborative world.

The phrase "connecting the dots" has become a popular idiom in our daily conversations. It refers to the act of finding and understanding the relationships between seemingly unrelated events or pieces of information. While it may sound simple, this concept is fundamental to problem-solving and even our search for meaning in life.

Navigating and Networking: Two Pilots' Guide to Building Strategic Relationships

Charting a cross-country flight in a single-engine Cessna is literally about connecting the dots, safely navigating from one airport to the next based on all kinds of data, including weather information, fuel supply, physics, timing, and mechanics.

In the vast expanse of the sky, a pilot's journey is meticulously mapped out, each waypoint carefully plotted to ensure a safe and efficient flight. Similarly, the art of networking, while seemingly boundless in its possibilities, demands a strategic approach, connecting dots and building relationships that propel one toward their goals.

This book was conceived during our 22-hour, cross-country flight from New York to Arizona in Toby's Cessna Skyhawk, with his friend and safety pilot Samir as his copilot. It is a testament to the power of networking, weaving together the experiences of two pilots—one a communications executive, the other a chief financial officer—who have shared the joy and responsibility of flying and have applied many of the principles of aviation to excelling in their lives and careers.

Nearly two decades ago, Toby and his husband Harlan Bratcher purchased their Cessna Skyhawk in Southern California, embarking on an unforgettable cross-country flight that marked the beginning of a lifelong friendship with their safety pilot and instructor, Samir Kanuga. Years later, as Toby, an experienced VFR-rated pilot, found himself flying once again with his friend and IFR-rated instructor Samir, the idea of capturing their combined expertise in networking and aviation took flight. You, dear reader, are holding the result in your hands. Think of this book as your flight manual.

Networking, like flight planning, requires careful consideration. Just as a pilot meticulously charts a course, navigating airspace regulations and weather patterns, individuals can build strategic connections by identifying goals, understanding the terrain of their industry, and being prepared to adapt to unforeseen circumstances.

At the heart of both aviation and networking lies the cornerstone of trust. In the cockpit, trust between the pilot and copilot is paramount. Similarly, networking thrives on trust by fostering open communication, mutual respect, and a willingness to collaborate.

Training plays a pivotal role in both aviation and networking. Pilots undergo rigorous training to master their craft, honing their skills in navigation, communication, and decision-making. Likewise, networking requires continuous learning, expanding one's knowledge base, refining communication skills, and developing the ability to connect with diverse individuals.

Discipline is the cornerstone of success in both aviation and networking. Pilots adhere to strict protocols, adhering to checklists and procedures to ensure safety and efficiency. Similarly, networking demands discipline, commitment to consistent engagement, nurturing relationships, and following through on commitments.

Adventure—the spirit of exploration—fuels both aviation and networking. Pilots relish the thrill of soaring through the skies and discovering new horizons. Similarly, networking opens doors to uncharted territories, leading to unexpected opportunities and collaborations often, really exciting new friends.

As the Cessna Skyhawk gracefully touched down in Arizona, your authors completed not only a cross-country flight but also the outline of this book, a practical guide for navigating one's human network.

This book is a flight manual for individuals seeking to build strategic connections to chart a course toward happiness and success in life.

Cleared for Takeoff

Have you ever wondered how some people seem to have a knack for networking? They know everyone, and they always seem to be in the right place at the right time. If you are like us, you have probably felt a bit envious. But do not worry; you are not alone. Networking is a skill that anyone can hone.

From the tips and techniques that have worked for us, we are going to share everything we know about networking. We will share some real-world tips and tricks from the best networkers in the world. Whether you're an introvert or an extrovert, whether you're just starting out in your career or you're a seasoned professional, this book has something for you.

Networking does not have to be intimidating. In this book, we show you how to network in a way that is comfortable for you and authentic.

We even delve into AI. In fact, we used AI in drafting this book. The thoughts, concepts, and words are our own. The images, references, corollaries, and metaphors are ours as well. The seemingly endless hours of editing are also all ours. But we used generative AI, including Bard, Google (now Gemini), and ChatGPT to help us mash up all the above before we fact-checked and edited the final book you are now reading.

If you're ready to take your networking skills to the next level, this book is for you. Let's get started. We are cleared for departure.

Developed in the mid-1880s and popularized by French artists Georges Seurat and Paul Signac, pointillism is a pivotal technique in art history. This painting style went on to be an important element in several artistic movements. Paintings created in the pointillist style—tiny dots that form to create a larger image—helped shape Impressionist, Neo-Impressionist, Post-Impressionist, and Fauvist works. From

Pissarro and van Gogh to Roy Lichtenstein and Damien Hirst's contemporary interpretations, the art of connecting the dots has resulted in incredible pointillist artwork.

Chapter 1: The Importance of Connections

The story of human networking stretches back millions of years; it is woven into the very fabric of our existence. Our earliest ancestors were nomadic hunter-gatherers who relied on intricate social networks for survival. These networks, composed of families, tribes, and bands, functioned as intricate webs of connection, fostering the sharing of resources and skills. Communication likely involved a combination of gestures, sounds, and the early seeds of language. Studies suggest that our ancestors weren't limited to connections within kinship lines.

Instead, they formed networks with individuals beyond immediate family, which played a crucial role in our evolutionary success. As civilizations arose, the need for communication transcended local boundaries and prompted the invention of writing and far-flung trade that ignited complex societies, which spurred on ever more sophisticated communication. Smoke signals and drums served as long-distance messaging systems while the rise of postal systems and early writing systems enabled information exchange over even greater distances. Trade routes, like the iconic Silk Road from China on the Pacific to Spain on the Atlantic, connected societies across continents. Cultural exchanges and the swapping of knowledge altered lives on a global scale.

The Industrial Age ushered in a revolution with the combustible engine and with transportation, drastically shrinking the space between communities. Trains, steamboats, and automobiles dramatically compressed distances, not only physically but also informationally.

Communication underwent another leap forward with the advent of the telegraph and telephone, allowing near-instantaneous connections across vast stretches of land. The Information Age was about to explode: radio, television, and the internet ushered in an era of mass communication and unprecedented global reach. Social media platforms like Facebook and Twitter (now X) represent the latest chapter in this evolving story, enabling individuals to interact—right now, this minute—anywhere in the world.

Throughout history, we can see that "networking" has served as a cornerstone of human survival and progress. Each era has witnessed technological advancements that revolutionized the way we connect, from smoke signals to social media. The nature of our networks has transformed from localized and face-to-face interactions to global and digital connections, presenting both opportunities and challenges. As we zip into the future, understanding this rich history will allow us to better navigate the ever-evolving landscape of human networking, harnessing its strengths and addressing its pitfalls.

Human connections are essential to our well-being. They provide us with love, support, and a sense of belonging. They also help us learn and grow, personally and professionally. In this chapter, we will discuss how to build and maintain strong connections.

Human connection isn't a luxury; it's a vital thread woven into the tapestry of our existence. Just as air nourishes our lungs, genuine connection with others fuels our emotional and physical well-being. It's a potent elixir that bolsters our spirit, shields us from adversity, and paints our lives with vibrant hues of meaning and purpose.

On a fundamental level, connection fosters a sense of belonging. We crave the feeling of being seen and accepted for who we truly are. In the embrace of genuine connection, we shed the isolating cloak of loneliness and find solace in one another. This sense of belonging fosters empathy and compassion, allowing us to weave a stronger social fabric where everyone feels valued and supported.

Beyond the emotional realm, human connection possesses the power to unlock a cascade of health benefits. Studies show that strong social ties bolster our immune system, which results in lower stress levels and contributes to a longer lifespan. The simple act of sharing a meal with a loved one or confiding in a friend can trigger the release of feel-good hormones, enhancing our overall sense of well-being. In essence, connection becomes a shield against negativity, a source of strength in hardship and a catalyst for a more fulfilling life.

So, in a world increasingly consumed by virtual interactions and fleeting connections, let us not forget the profound significance of genuine human touch. Let us nurture our existing relationships, forge new bonds, and celebrate the simple joy of being connected. For in the tapestry of human connection lies the true essence of our humanity, a vibrant testament to the power of love.

Strong human connections have a number of benefits, including:

- Improved mental and physical health. Studies have shown that people with strong social connections have lower rates of depression, anxiety, and loneliness. They also tend to live longer and healthier lives.
- Greater happiness and satisfaction. People with strong social connections tend to be happier and more satisfied with their lives. They ~~also~~ have higher self-esteem and resilience.
- Increased opportunities. Strong social connections ~~can~~ lead to new opportunities in all areas of life— work, school, and personal relationships.
- Greater support and understanding. Strong social connections bolster us during difficult times. They help us to maintain and sustain.

Professional connections that we have with colleagues, mentors, and other professionals in our field can be essential for our career success, providing us with:

- Mentorship and guidance. Professional connections provide mentorship and guidance as we navigate our careers.
- Job opportunities. Professional connections lead to new job opportunities.

- Information and resources. Professional connections provide information and resources.
- A sense of community. Professional connections give us a sense of community within our field.

Personal connections are essential for our emotional and psychological well-being, providing us with:

- Love and support. Our personal connections provide us with the love and support that we need to thrive.
- A sense of belonging. Our personal connections give us a sense of belonging and acceptance.
- Companionship. Our personal connections provide us with companionship and enjoyment.
- Opportunities to learn and grow. Our personal connections help us to learn and grow as individuals.

Personal connections can be divided into different types:

- Close connections: These are connections we have with the people we are closest to— family and close friends.
- Casual connections: These are connections we have with people we know less well – acquaintances and coworkers.
- Online connections: These are connections we may never meet face to face through social media and other online platforms.

Each type of connection has unique benefits. Close connections provide us with emotional support and a sense of belonging. Casual connections help us expand our social circle and learn new things. Online connections help us stay connected with friends and family who live far away as well as with strangers interested in what we are interested in.

Community connections can be divided into different types:

- Formal connections: These are connections we make through organizations and institutions—religious groups, civic organizations, professional associations.
- Informal connections: These are connections we make through social groups and activities— sports teams, book clubs, volunteer groups.

Formal connections help us make a difference in our community and network with people who share our interests. Informal connections help us laugh, build relationships with neighbors, and create a sense of community.

No matter what type of connection it is, all connections are important. Connections help us live happy, fulfilling lives. They also help us make a difference in the world.

Degrees or Strength of Connection

Within each of these broad categories are many types of connections. For example, professional connections include:

- Direct connections: These are people we know and interact with on a regular basis – colleagues, teammates, clients.

- Indirect connections: These are people we know through other people— for example, a friend of a friend or a colleague of a former colleague.
- Weak ties: These are connections ~~that~~ we have with people we don't know well or have not met— networking events or online.

Weak ties can be especially important for networking. They can often help us reach new people and discover new opportunities.

By cultivating strong connections across all aspects of our lives, we improve our mental and physical health, increase happiness and satisfaction, and create opportunities for ourselves and others.

We Never Fly Solo in Life

Piloting a plane, even solo, can be a powerful metaphor for being connected to other people in several ways, thanks to interactions with air traffic controllers, Universal Communications stations (UNICOM), and radar.

A pilot navigating in the vast sky isn't truly alone but constantly guided by a network of unseen voices and technology. Air traffic controllers are guardians whose job is to orchestrate the flow of planes; they inform all those pilots up there how to navigate safely.

This mirrors how we rely on unseen social structures, routines, and communication to navigate the complexities of society.

While a pilot may sit alone in the cockpit, they're reliant on the efforts of many others—engineers who built the plane, the ground crew who prepare it, air traffic controllers, and fellow pilots sharing the airspace. This interdependence reflects the reality that our individual success and well-being are interwoven with others' contributions.

UNICOM channels act as bridges between individual pilots, allowing them to share information, report conditions, and, if necessary, coordinate actions. This echoes how clear and timely communication is vital for building trust and collaboration in any interpersonal relationship or societal interaction.

Radar becomes an extension of the pilot's senses, painting a picture of the surrounding airspace. This parallels how empathy and social awareness allow us to perceive and understand the perspectives and needs of others and they of us, fostering connection and cooperation—a two-way street.

As one can see, in the vastness of the sky, a pilot's safety hinges on trusting the guidance and information received from others. This vulnerability mirrors the courage it takes to open up and trust others in everyday life.

Piloting a plane, even solo, highlights the invisible threads that connect us all. It reminds us that even in moments of apparent isolation, we navigate life within a web of interdependence. By recognizing and embracing these connections, we build a more supportive and cooperative world.

Discussion Questions

Each chapter of the book includes discussion questions at the end to help readers practice what they have learned. For Chapter 1, consider these questions:

- Which benefits have you realized in having strong personal and/or professional relationships?
- Reflect on a time when your outreach backfired. Why? What would you do differently?
- How have you developed long-term relationships in your life so far? Are there specific ones that you targeted and built? Why and how?

The Millennial Job Interview -
https://www.youtube.com/watch?v=Uo0KjdDJr1c

The video "The Millennial Job Interview" is a comedic sketch that highlights the stereotypical traits of Millennials in a workplace setting. It portrays an exaggerated job interview where a young Millennial candidate displays a sense of entitlement, lack of real-world experience, and an over-reliance on technology and social media. The sketch humorously emphasizes the generational gap and the misunderstandings that can arise between Millennials and their potential employers, making a broader point about the evolving expectations and dynamics in modern job markets.

Chapter 2: Intergenerational Views on Networking

Networking is the process of building and maintaining relationships with people who can help you achieve your goals. It is a powerful tool.

Different generations have different views on networking. This is due to a number of factors, including generational differences in communication styles, work values, expectations, and even fear of change.

For the first time in history, thanks to longer life expectancies, there are up to four generations in our workplaces:

- Traditionalists (builders also referred to as The Silent Generation): Born 1925–1945
- Baby boomers: Born 1946–1964
- Generation Xers (Gen Xers): Born 1965–1980
- Millennials (Gen Yers): Born 1981–1996
- Generation Zers (Gen Zers): Born 1997–2009
- Generation Alpha (Gen Alpha): Born 2010 or later

Traditionalists 'Builders' Views on Networking (Born 1925–1945)

The Traditionalist group is nearing the end of life and tends to be more set in their ways. Traditionalists have completed their careers and settled down. They are not immune to networking, for example, as Toby's stepfather, Neil Vogler, moved into a retirement community at 77 and for the last five years of his life made some of his closest and most beloved friends via shared dinners and social activities. Humans are social animals, and Neil knew himself, understood what he cared about, and was proactive in engaging others. In that sense, Neil optimized his networking as he grew older.

Traditionalists like Neil are also known as Builders or the Silent Generation (1925-1945). They make up the oldest living generation growing up during the Great Depression and World War II. They are known for their resilience, work ethic, and patriotism. They are the first generation to grow up with television, and they witnessed major technological and cultural changes. Today, the Traditionalists are mostly retired but remain active in their communities, continuing to make contributions to society.

Some notable Traditionalists include the late Queen Elizabeth II (born in 1926) and the "Wizard of Omaha," investor Warren Buffett (born in 1930).

Queen Elizabeth II worked until the very last day of her life, and her legacy is one of enduring stability and evolving modernity. As the longest-reigning British monarch, she witnessed and navigated a period of immense change, from the twilight of the British empire to the rise of the internet. Her unwavering presence provided a sense of continuity amidst social upheaval, political shifts, and global turmoil. She played a pivotal role in the monarchy's own adaptation, embracing public engagement and modernization while upholding cherished traditions. While her association with the colonial past sparked complex conversation, her efforts to foster cooperation and collaboration within the Commonwealth—a diverse grouping of former colonies—cannot be ignored. Ultimately, Queen Elizabeth II's legacy is one of resilience, adaptation, and a commitment to service, leaving an indelible mark on both Britain and the world stage.

Warren Buffett's legacy transcends mere wealth, etching itself onto the very definition of investment and philanthropy. His value-investing philosophy, emphasizing long-term ownership and intrinsic value, has empowered generations of investors, democratizing wealth creation and challenging market fads. Berkshire Hathaway, his investment behemoth, stands as a testament to shrewd vision, weathering economic storms and generating staggering returns. Yet, Buffett's impact extends beyond financial acumen. His commitment to giving, pledging over $50 billion to charitable causes, has redefined the role of billionaires in society, inspiring others to leverage their fortunes for the greater good. His down-to-earth demeanor and folksy wisdom have also endeared him to the masses, making complex financial concepts accessible and demystifying the world of high finance. In a world obsessed with quick gains and fleeting trends, Warren Buffett's legacy stands as a beacon of goodwill, patience, discipline, and a profound belief in the power of human connection. He is not just a financial titan, but a cultural icon, a testament to the enduring value of wisdom, humility, and a long-term perspective on life and wealth.

Baby Boomer Views on Networking (Born 1946–1964)

Toby, co-author of this book, and his fellow baby boomers are the generation born between 1946 and 1964. They are the largest generation in the workforce and are nearing retirement age. They tend to have a traditional view of networking and believe in building relationships face-to-face, valuing personal connection and trust.

Members of the baby boomer generation are renowned for their robust work ethic and goal-oriented nature. They exhibit a strong commitment to hard work and place high value on face-to-face interactions. Despite entering the workforce before the advent of certain technologies, baby boomers have adapted to incorporate new skills, including those related to computers, into their professional lives.

Engaging Baby Boomers

Baby boomers may feel more at ease with traditional recruitment processes involving formal résumés and face-to-face interviews. Their job searches often rely on advertisements, word of mouth, and referrals. Effective retention strategies for this generation emphasize offering meaningful work with schedules that accommodate their needs. Mentorship programs tap into their wealth of knowledge, allowing them to share their extensive experience with colleagues and newbies.

Thriving on the opportunity to engage in impactful projects that shape the future of the company or society. They take pride in their work and derive satisfaction from-recognition – public acknowledgment, professional development opportunities, chances to prove themselves, and special perks like an office, title, or parking spot. Retirement benefits, including phased retirement, part-time schedules, and retirement/financial counseling, are also valued, reflecting their preference for retirement security.

Baby boomers seek recognition for their extensive experience and skills, making them valuable mentors. Good managers know enough to leverage these traits, encouraging boomers to share their industry knowledge with younger employees.

As many baby boomers approach retirement, they appreciate flexible work policies. Offering reduced schedules, the option to work from home, or alternative hours can influence experienced staff members to extend their careers and enhance the smooth transition for new employees. Health care and retirement benefits, including a 401(k) match, are highly desirable to this generation.

Baby boomers are more likely to network through familiar people and organizations, such as professional associations and industry events. They are also more likely to rely on mentors and sponsors.

Baby boomers are the largest generation in American history. They grew up during a period of economic prosperity and social change, and they are known for their optimism, idealism, and activism. They were the first generation to go to college en masse, and they helped shape the counterculture movement of the 1960s and 1970s. Today, despite reaching retirement age, they remain a powerful force in American politics and culture.

Some notable individuals from the baby boomer generation include Bill Clinton (born in 1946), Steve Jobs (born in 1955), Oprah Winfrey (born in 1954), and Jodie Foster (born in 1962).

Bill Clinton's legacy remains a complex mosaic, shimmering with economic triumphs like the longest peacetime expansion in US history, yet stained by personal scandals like the Monica Lewinsky affair and his impeachment. He championed globalization through trade deals like NAFTA, signed welfare reform, and pushed for LGBTQ+ rights through "Don't Ask, Don't Tell." While hailed as a "New Democrat" who revitalized the party, his legacy is also critiqued for growing income inequality and expanding mass incarceration. Ultimately, Clinton's impact remains an ongoing debate, etched in both economic prosperity and political controversy.

Steve Jobs's legacy shimmers like a bitten apple, a complex blend of innovation, sleek design, and an insatiable hunger for perfection. He wasn't just a tech wizard but also a maestro of minimalism, morphing clunky computers into objects of desire like the iconic iMac and iPhone. His product launches weren't just spec-fests, but choreographed experiences that ignited passion and loyalty. He wasn't just selling gadgets, but a future when technology seamlessly wove into our lives, amplifying creativity and connection. Jobs, as demanding and flawed as he was, dared to dream audaciously, pushing boundaries and forever altering how we interact with the world around us. His bite-sized legacy is a testament to the power of human potential, a reminder that even the most impossible dreams can bloom when nurtured with vision, relentless pursuit.

Oprah's legacy is a vibrant tapestry woven from threads of shattered barriers, empowered voices, and boundless giving. As a media icon, she redefined the landscape, tackling taboo topics and offering a platform for marginalized voices. Beyond entertainment, she's a philanthropic powerhouse, empowering girls, aiding communities, and inspiring generations with her unwavering belief in the power of empathy and self-discovery. Oprah's impact transcends mere entertainment; her story is a testament to the enduring power of a single voice to touch countless lives and leave a legacy that continues to empower.

Generation X Views on Networking (Born 1965–1980)

Generation X (Gen X) is the generation born between 1965 and 1980. They are the middle generation in the workforce and are known for their work-life balance. Caught between the baby boomers and millennials, Gen Xers emerged during the rise of personal computers. Generally, this generation is more educated than its predecessors. Regarded as self-reliant and industrious, Gen Xers are often seen as financially prudent.

Engaging Gen Xers

Gen Xers are adept at using technology and online recruitment tools, yet they also value face-to-face interactions. Achieving the right blend of benefits is vital for attracting and retaining Gen X talent. Collaboration in multigenerational teams can be challenging. Teams primarily composed of baby boomers and millennials tend to emphasize collaboration while Gen Xers may be more inclined to work independently and present individual research to the team.

Gen Xers view jobs as contractual agreements, placing practicality at the forefront. They expect fair compensation and opportunities for additional earnings. Skill-building and credentialing opportunities are highly valued, and Gen Xers seek a balance through time off. Recognition for this group includes cash and non-cash awards, professional development, work-life balance options, and the freedom to approach tasks in their own way.

In the workplace, Gen Xers may favor individual-centric environments with flexibility in managing workloads and greater physical and psychological space. Having adapted well to remote work during the pandemic, this generation, transitioning from parenting to potential caregiving roles, places a premium on job flexibility.

Gen Xers, well-established in their careers, bring valuable experience that managers should appreciate. They prefer less supervision and greater autonomy in completing responsibilities, frequently desiring schedules allowing for a healthy work-life balance. Often raising families, they prioritize health care coverage, flexible work arrangements, on-site daycare, and other perks supporting work-life balance. Gen Xers particularly value monetary benefits such as stock options, dental and vision coverage, 401(k) and retirement savings plans, as well as financial planning services.

Generation X tends to have a more informal view of networking. They are comfortable networking online and through social media. They also value authenticity and transparency. Thus, they are more likely to network through personal connections and work colleagues, often relying on peers for support and advice.

Generation X is known for its independence, self-reliance, and skepticism. They grew up during a time of economic recession and social upheaval, and they witnessed the rise of the personal computer and the internet. They are often described as the "latchkey generation" because they came of age during a time when both parents worked outside the home. Today, Gen Xers are in the prime of their careers, making significant contributions to the workforce and society.

Some notable individuals from the Gen X generation include Justin Trudeau (born in 1971), Elon Musk (born in 1971), and Jennifer Lopez (born in 1969).

Justin Trudeau's legacy is still being written, a complex tapestry woven with achievements and controversies. His supporters hail him as a champion of social progress, citing his commitment to gender equality, LGBTQ+ rights, and indigenous reconciliation. They point to his efforts on climate change, including a national carbon tax, and his ambitious plan for early childhood education and childcare. His detractors criticize his perceived political pragmatism, accusing him of broken promises and missed opportunities. They highlight scandals like the SNC-Lavalin affair and the WE Charity controversy, questioning his judgment and ethical conduct.

Beyond policy, Trudeau's image and networking skills remain a defining element. His charisma and fluency on social media helped propel him to power, but his "sunny ways" sometimes feel superficial. His extensive network, built from his time as a teacher, MP, and global advocate, has opened doors and secured partnerships, but it also raises concerns about potential conflicts of interest and undue influence. Whether Trudeau's legacy will be one of transformative change or political expediency remains to be seen, but his ability to connect with voters and navigate complex political landscapes will undoubtedly be a part of the story.

Elon Musk's legacy is a maelstrom of achievements and controversies, driven by his audacious vision and unconventional approach. Spearheading revolutionary companies like Tesla and SpaceX, he's a pioneer in electric vehicles and space exploration, yet his brash tweets and penchant for disruption leave a trail of admirers and detractors. His networking, though, is undeniable: a blend of Silicon Valley connections, childhood friendships, and online savvy, amplifying his influence while raising concerns about manipulation and echo chambers. Whether lauded as a visionary or criticized as a chaotic force, Musk's legacy remains as unpredictable as his rockets, fueled by a relentless pursuit of the impossible.

Jennifer Lopez's legacy glitters like the diamonds on her mic, a multifaceted masterpiece of Latina power, relentless hustle, and unwavering authenticity. From Bronx dance floors to Hollywood stages, she's defied stereotypes, built empires across music, film, and fashion, and empowered countless Latinas to dream bigger. Her networking? A masterclass in strategic charm and genuine connection. Years of building relationships, from dancers to directors, have woven a web of trust and respect, opening doors, yet Lopez remains fiercely loyal to her roots. Lopez's legacy reminds us that success isn't just about talent; it's about building bridges, lifting others, and proving that "Jenny from the Block" can conquer the world, one dazzling step at a time.

Millennials' (Gen Yers') Views on Networking (Born 1981–1995)

Samir and his fellow millennials are the generation born between 1981 and 1995. They are the largest generation in the workforce, and they are known for their digital fluency.

Many millennials entered the job market during a recession, significantly shaping their perspectives on long-term careers. Growing up in a society transformed by the internet, they exhibit a greater comfort with digital communication compared to earlier generations. In the workplace, millennials often opt for efficient means of communication such as instant messages, emails, or texts, rather than engaging in face-to-face conversations. Despite this digital preference, they highly value feedback from managers and actively seek guidance from mentors.

Engaging Millennials

Millennial job candidates typically anticipate a technology-driven application process, encompassing mobile-optimized applicant tracking systems, applications integrating with platforms like LinkedIn, and learning about career opportunities through social recruiting. Retention strategies should emphasize the development of a skill-focused training program addressing their aspirations for leadership training, skills enhancement, and career advancement.

For millennials, a stimulating and enjoyable work environment is paramount. Placing this group in collaborative work teams, fostering positive relationships with managers, and involving them in decision-making processes are key motivational factors. They find meaning in their work and appreciate regular feedback, contributing to their self-confidence and proactive attitude. They are inclined toward ongoing training and skill acquisition that enhances their résumés. Listening to their ideas and opinions is crucial, given their expectations of being heard, instilled by their upbringing. Millennials are motivated by fun and stimulation. Employers can maintain lower turnover rates and higher productivity by offering rewards such as cash and non-cash awards, work-life balance options, the freedom to approach tasks uniquely, opportunities for internal growth, and recognition.

Performance quality holds significant importance for millennials, who evaluate their managers based on the substance of their work rather than traditional measures like hours spent in the office. Transparent and honest communication is essential when interacting with millennials, fostering an environment in which questions are welcomed.

Millennials prioritize career development opportunities and benefits that support a healthy work-life balance. Examples include career development programs, affordable health insurance, on-site daycare facilities, mortgage services, 401(k) and retirement planning, and generous paid time off (PTO).

Millennials have a very different view of networking than previous generations. They are comfortable networking online and through social media and value diversity and inclusion. They are also more likely to rely on informal mentors and coaches.

Millennials are the most diverse generation in American history. They grew up during a time of rapid technological and cultural change, and they are known for their entrepreneurial spirit, their social media savvy, and their commitment to social justice. They are the first generation to come of age in the digital age, and they are comfortable using technology to connect with others, learn new things, and make a difference in the world. Today, millennials are entering the workforce and starting families. They are poised to have a major impact on America's future.

Notable millennials include Mark Zuckerberg (born in 1984), Alexandria Ocasio-Cortez (born in 1989), and Zhang Yiming (born in 1983).

Mark Zuckerberg's legacy is a double-edged sword forged in the fires of innovation but tempered by controversy. He revolutionized social connection with Facebook, democratizing information and fostering community, yet he faces accusations of data privacy violations and enabling the spread of misinformation. His networking, a master class in blending youthful charm with Silicon Valley prowess, catapulted him to the top but raised concerns about wielding undue influence and prioritizing growth over ethics. Whether his name is etched in history as a visionary architect of connection or a cautionary tale of unchecked power remains to be seen, but his impact on our digital lives is undeniable.

Alexandria Ocasio-Cortez's legacy is a still-unfolding tapestry woven with threads of fiery activism, bold policy proposals like the Green New Deal, and a mastery of social media that redefined political engagement. Her approach to networking is as unconventional as her rise: bypassing traditional power structures, she leverages online communities and grassroots movements to build a diverse coalition of young progressives, injecting fresh voices into the political landscape. Whether her legacy will be one of transformative change or political disruption has not yet been determined, but AOC's unapologetic voice and innovative networking have undoubtedly shaken up the political establishment and ignited a generation.

Zhang Yiming, the enigmatic founder of TikTok, has carved a unique path in the tech world. His legacy is still being written, but it's already a fascinating blend of innovation, virality, and controversy. On the one hand, Yiming is hailed as a visionary who democratized entertainment, providing a global stage for anyone with a smartphone and a creative spark. His understanding of virality and the power of short-form content revolutionized how people create and consume media.

Unlike many tech CEOs who court media attention and cultivate public personas, Yiming operates in the shadows. He rarely gives interviews, avoids public appearances, and maintains a meticulously crafted online presence. This calculated aloofness has fueled speculation and intrigue, adding to the mystique of the man behind the viral behemoth.

Generation Z Views on Networking (Born 1995–2009)

Generation Z is the generation born after 1996. They are the youngest generation in the workforce and are known for their social consciousness and their entrepreneurial spirit.

Members of Generation Z are true digital natives, considering smartphones and other devices indispensable elements of their lives. Unlike previous generations, they tend to focus more on personal qualities like humor, wit, and intelligence rather than factors such as race or ethnicity. This shift in perspective is largely attributed to the impact of technology on their interpersonal relationships.

Engaging Gen Zers

To effectively attract and manage Gen Z in the workplace, employers should establish a robust digital presence. Given Gen Zers' reliance on the internet and social media for researching potential employers, a tech-driven and efficient hiring process is preferable. Gen Z employees thrive when provided access to cutting-edge technology. They are eager to kickstart their careers and often prioritize salary over benefits.

Having witnessed the aftermath of the 2007–08 financial crisis through their parents, job security is a key concern. Generation Zers seek somewhat stable opportunities and typically remain with the same company for two to four years. They value flexibility, input on process improvements, flexible work hours, access to innovative tools, remote work options, and environments that prioritize social responsibility and diversity.

Highly collaborative management is a defining characteristic of Generation Z. They look to management to establish a strong mission, provide clear expectations, offer regular constructive feedback, and set an example for continuous learning and growth. Companies should focus on attracting the right talent, investing in development, and creating mentoring, coaching, and learning opportunities with senior staff when formulating management policies for this generation.

Workplace flexibility takes precedence as one of the most sought-after benefits for Gen Z employees, alongside health care and professional development. Other priorities include student debt assistance, competitive salaries, financial incentives such as raises after project completion, tuition reimbursement, formal training, mental health benefits, wellness programs, commuter benefits, and parental leave.

Generation Z has a very fluid view of networking. They are comfortable networking online and through social media, and they value authenticity and purpose. They are more likely to network through personal connections and social causes. They are also more likely to rely on peers and mentors for support and guidance.

Generation Z is the most technologically savvy generation to date. They grew up with the internet, smartphones, and social media at their fingertips. They are known for their entrepreneurial spirit, social justice activism, and commitment to mental health awareness. Gen Zers are just starting to enter the workforce and college, but they are already making their voices heard on important social and political issues.
Notable Gen Zers include Shawn Mendes (born in 1998) and Greta Thunberg (born in 2003).

Shawn Mendes's legacy, like his melodies, is a blend of youthful charm and soulful maturity. Bursting onto the scene as a teenage heartthrob, he captivated audiences with his raw vocals and relatable lyrics. His early success, fueled by social media savvy and strategic collaborations, cemented him as a pop force, but it also raised questions about longevity and artistic growth.

His approach to networking reflects this duality. While he's cultivated relationships with established stars, Mendes has also actively championed emerging artists, fostering a sense of community within the industry. He leverages his platform to advocate for social causes close to his heart, showcasing genuine empathy alongside his pop-star persona. This balance between ambition and authenticity has resonated with fans, creating a loyal following extending beyond the typical pop demographic.

Mendes's legacy, though still evolving, is one of bridging generations and musical styles. He's mastered the art of playing the pop game while staying true to his artistic vision, inspiring young musicians and proving that sincerity can coexist with stadium-filling success. Whether his future holds chart-topping anthems or more introspective explorations, one thing is certain: Shawn Mendes's melody lingers long after the final note, leaving a mark on the pop landscape and reminding us that, sometimes, the most powerful network is built on genuine connection.

Greta Thunberg's legacy is a still-unfolding storm, a whirlwind of youthful activism that has swept across the globe, challenging apathy and demanding action on climate change. Her voice, amplified by social media and a strategic approach to networking, has resonated with millions, uniting a generation in the fight for a sustainable future.
Unlike many activists who rely on traditional channels of influence, Thunberg has harnessed the power of the internet to connect directly with her audience. Her viral #FridaysForFuture movement, born from solitary school strikes outside the Swedish parliament, mobilized millions of students worldwide, demonstrating the power of grassroots organizing in the digital age.

Thunberg's approach to networking extends beyond social media. She has strategically built alliances with scientists, policymakers, and even other youth activists, forging a diverse coalition united by a common cause. Her willingness to engage in open dialogue, even with her detractors, has earned her respect and admiration from across the political spectrum.

Whether Thunberg's legacy will be one of transformative change or continued frustration remains to be seen, but one thing is certain: She has redefined the role of the activist in the 21st century, proving that a single voice, amplified by the right network, can shake the world's foundations.

Thunberg's story is a testament to the power of conviction, the importance of strategic communication, and the boundless potential of youth activism in the digital age. She is already a force to be reckoned with, a storm that will continue to rage until the fires of climate change are extinguished.

Generation Alpha's (Gen Alpha's) Views on Networking (Born 2010 or Later)

Generation Alpha is the youngest generation today. They are growing up in a rapidly changing world, facing challenges such as climate change, economic inequality, and social media addiction. They are known for their creativity, resilience, and global outlook. Gen Alphas are still very young, but they have the potential to shape the future of the world.

One study by the Pew Research Center found that 93% of children aged three to five have access to smartphones or tablets at home. This is up from 75% in 2011. The study also found that children aged six to eight are more likely to use technology than watch television.

These findings suggest that Generation Alpha is growing up in a world in which technology is ubiquitous and integrated into their daily lives from a very early age. As this generation matures and takes on more responsibilities, it will be interesting to see what impact their deep connection to technology has on society.

Engaging Generation Alpha without Infringing on Parental Gateways

Generation Alpha easily navigates the online world, consuming content through a kaleidoscope of platforms and devices. Engaging them, however, poses a unique challenge—their young age necessitates parental oversight, so how do we create captivating experiences for Generation Alpha while respecting the role of their guardians?

- Prioritize interactive and visual content: Generation Alpha thrives on stimulating formats. Ditch dry text walls for interactive games, quizzes, polls, and AR/VR experiences. Make visual storytelling your mantra, employing vibrant graphics, catchy animations, and short-form videos to grab their fleeting attention.

Think TikTok, Instagram Reels, and bite-sized YouTube formats.
- Partner with parents; don't bypass them: Don't view parents as gatekeepers—embrace them as collaborators. Develop content that appeals to both generations, sparking family conversations and shared enjoyment. Consider co-creation platforms where parents and children can work together on digital projects, fostering parent-child bonding and brand engagement.
- Transparency and trust are key: Generation Alpha is remarkably conscious of data privacy and ethical practices. Be transparent about how you collect and use their information, prioritizing data security and parental control tools. Building trust early on is crucial for long-term loyalty.
- Champion their values: This generation is passionate about social and environmental issues. Align your brand with its values, addressing topics like sustainability, inclusivity, and mental health in your content. Let them know you share their concerns and are actively working toward solutions.
- Leverage micro-influencers: Forget celebrity endorsements. Gen Alpha trusts their peers far more than traditional stars. Partner with relatable micro-influencers who resonate with their specific interests and communities. Collaborate on authentic content that feels organic and relatable, not forced advertising.
- Make learning fun and accessible: Remember, education and entertainment are intertwined for Gen

Alpha. Design content that blends knowledge and fun, making learning an engaging, interactive experience. Educational games, AR/VR simulations, and gamified learning apps are fertile ground for engagement and knowledge transfer.
- Think phygital experiences: Don't confine your engagement to the digital realm. Bridge the gap between online and offline with phygital experiences connecting Gen Alpha's virtual and physical worlds. Interactive events, AR-enhanced scavenger hunts, and gamified product installations create lasting memories and brand associations.

A Word About Phygital

Phygital is a portmanteau of *physical* and *digital*, referring to blending physical and digital experiences. It's about seamlessly integrating the best of both worlds to create engaging and enriching user experiences.

Think of it as a bridge between the two worlds, removing the lines that traditionally separated them. Instead of siloed, isolated experiences, phygital aims to create a cohesive ecosystem in which the physical and digital aspects enhance and amplify each other.

Here are some key characteristics of phygital experiences:

- Integrated technologies: Digital technologies are embedded into the physical environment, such as AR displays in stores, interactive kiosks, or gamified product trials.
- Seamless transitions: Moving between physical and digital interactions should be effortless and intuitive, creating a smooth user journey.
- Personalized experiences: Phygital experiences often use data and personalization technologies to tailor the experience to individual users' preferences and interests.
- Focus on engagement: Engage users, sparking curiosity, interaction, and participation.
- Value creation: Phygital should add value to the user experience, offering something beyond what either the physical or digital alone could provide.

Here are some real-world examples of phygital experiences:

- Amazon Go stores: Shoppers walk in, grab what they need, and simply walk out, with payment automatically charged to their account.
- Interactive museum exhibits: AR overlays on physical exhibits provide additional information and interactive elements.
- Gamified retail experiences: Points and rewards are earned for interacting with products or completing tasks in-store.

- Virtual try-on technology: Shoppers can see how clothes or products would look on them through AR or VR.
- Phygital fitness challenges: Blending online tracking with real-world events and activities motivates participants.

The possibilities for phygital experiences are endless, with their importance growing as technology and consumer behavior evolve. By merging the physical and digital worlds, phygital has the potential to revolutionize how we interact with brands, learn, shop, and experience the world around us.

Remember, engaging Generation Alpha isn't about simply targeting them directly. It's about building trust with both them and their parents, creating content that sparks entertainment and learning, and aligning your brand with their values. By respecting their young age and incorporating their parents into the engagement journey, you can unlock a world of possibilities for meaningful connection and long-term loyalty.

Notable Gen Alphas include the "Corn Kid" (born in 2015) and Princess Charlotte of Wales (born in 2015)

As the "Corn Kid," Tariq's seven-year-old's viral fame still echoes across the internet, a testament to pure joy and honest enthusiasm in the digital age. His legacy, though young, is one of unadulterated passion for a simple thing: corn. His infectious, "It's corn!" chant and genuine excitement for the subject turned a seemingly mundane interview into a cultural phenomenon.

The CEO of Corn | Recess Therapy #cornboy

https://www.youtube.com/watch?v=1VbZE6YhjKk

Tariq's networking isn't orchestrated or strategic; it's built on sheer enthusiasm and genuine connection. He's embraced the spotlight, appearing on talk shows, attending events, and collaborating with other internet stars. His interactions are marked by childlike wonder and open curiosity, forging unexpected connections with everyone from musicians to celebrities.

This organic approach to networking has endeared him to audiences, refreshing the often-calculated world of online fame. He reminds us that, sometimes, the most powerful connections are built on unfiltered authenticity and a shared love for, well, corn. Tariq's legacy is a reminder that joy and passion, even for the most unexpected things, can connect us across generations and platforms, leaving a sweet, corny aftertaste in the digital landscape.

Princess Charlotte of Wales's legacy, at seven, is not yet etched in stone. However, her short life so far hints at a potential for a future filled with grace, influence, and a modern approach to royal engagement.

While her official networking is limited by her age and royal protocol, Charlotte's charm and charisma shine through in glimpses. Her confident interactions with crowds, playful spirit with her siblings, and genuine interest in others suggest a budding ability to connect and build relationships. She's already embraced social media through her parents' platforms, subtly showcasing potential for navigating the digital world with the same ease with which she navigates palaces.

Charlotte's future networking, however, might differ from the traditional royal model. Her generation, raised on technology and social awareness, might see her leverage technology such as video calls and virtual appearances to connect on a global scale, bypassing the limitations of a physical presence. She could be a champion for causes close to her heart, using her platform to advocate and amplify voices to resonate with a younger generation.

Of course, this is all speculation. Princess Charlotte's path is still unfolding, a blank canvas waiting to be filled with her unique experiences and choices. But if her early years are any indication, she might forge a legacy of genuine connection, digital savvy, and a modern approach to networking that could redefine the very meaning of royal influence.

It is important to note that these are just generalizations about each generation. There is a great deal of variation within each generation, and individuals should not be stereotyped.

Bridging Leadership

With so many generations working and socializing together today, we believe the optimal way forward is via bridging leadership—trying to find common ground with others versus focusing on differences.

This advertisement in early 2024 from the Australian/New Zealand Lamb Association created a fun message on how food can bring generations together: https://www.tiktok.com/t/ZT8VfQF7h/.

By humorously introducing different generations through their differences, and then drawing them together over a barbecue of lamb, an Aussie favorite, the generations bond over their shared affinity for lamb and barbecues.

Bridging the Gaps via Shared Passions

In a world increasingly divided by headlines and algorithms, finding common ground feels like an Olympic sport. But fear not; there's an antidote to this fragmentation: bridging leadership. Forget forceful decrees and hierarchical directives; bridging leadership thrives on a more fundamental force—shared passions.

Imagine a diverse team: A Gen Z tech whiz, a baby boomer baker, a Gen X architect, and a millennial musician. Individually, their interests seem worlds apart. Yet, begin a conversation about soccer (or football, depending on who you ask), and suddenly sparks fly. The tech whiz analyzes player stats, the baker reminisces about Maradona's magic, the architect dissects stadium design, and the musician hums an iconic World Cup anthem. The gap between generations evaporates, replaced by a shared love of the beautiful game.

This is the power of bridging leadership. It recognizes that passions, like food, music, art, or even the furry residents of our homes, transcend age, background, and belief. A leader who champions these shared interests becomes a conductor, harmonizing the unique notes of individual experiences into a beautiful collective melody.

Consider a community garden where a retired accountant mentors a shy teenager through growing tomatoes. Or a neighborhood choir in which a wheelchair user and a high school athlete find their voices blending seamlessly in a Mozart aria. These seemingly disparate individuals, brought together by the love of earth or music, form a tapestry far richer than any single thread.

Bridging leaders don't force unity; they nurture it. They create safe spaces for honest discussions about recipes, paint strokes, or the purrfect cat breeds. They listen, encourage, and celebrate the diverse perspectives that enrich these conversations. In doing so, they chip away at the walls of prejudice and misunderstanding, brick by brick.

The benefits extend far beyond shared laughs over a potluck or a spirited debate about the greatest rock band. Bridging leadership fosters empathy, respect, and collaboration. It teaches us to see the world through different lenses, expanding our understanding and enriching our lives.

So, the next time you find yourself surrounded by seemingly different people, remember the power of shared passions. Look for the gleam in people's eyes when they talk about their favorite dish, tap their feet in rhythm to a familiar beat, or shower love on their furry companions. There, in the fertile ground of shared interests, lies the potential for bridging divides and building a more harmonious world, one passion at a time.

And, who knows, you might just discover a hidden passion in yourself, forging a new connection with someone you least expected. The world of shared experiences awaits, so let's bridge the gaps—together.

Discussion Questions

For Chapter 2, consider these questions:

- Reflect on an intergenerational experience in your workplace. What went well, and what went off the rails? Why?
- Hypothesize the ideal generational mix of employees to hire for a new clothing store? What about a tech start-up? Describe your ideal team of talent and how you would source them.
- Apple and Google are two companies that actively hire older generations as part of their commitment to DEI (diversity, equity, and inclusion). What do you see as the pros and cons of their approach?

Chapter 3: Mixing: Connecting with Anyone

Today's modern workforce is undergoing a profound transformation with the emergence of intergenerational teams. In today's dynamic business environment, organizations are increasingly recognizing the value of diverse perspectives that span different age groups. Intergenerational teams, composed of individuals from baby boomers to Generation Z, bring a wealth of experiences, skills, and insights. However, along with opportunities, these diverse teams also face unique challenges that require strategic navigation.

Whomever the universe sends into our lives, we recommend welcoming them to our mix. By assuming a positive attitude of adventure and curiosity when encountering new or different people, you will build your network and legacy more effectively. Enjoy mixing it up. Strive to mix it up.

Every day presents us with new and different people as well as those whom we already know and with whom we interact. Think of your own workplace and an example of a time when you were left scratching your head over a statement or act from a colleague of a different generation. Did you try to equate that person to a relative of that same generation (e.g., "That sounds just like something my dad would say" or "That's exactly how my daughter is.")? It is important to take stock of the dynamics, challenges, and opportunities inherent in intergenerational teams.

Navigating Life's Airspace: Communicating with All Incoming Aircraft

Imagine yourself alone in your cockpit soaring through the clouds. You're in control, yes, but truly navigating the vastness requires more than just piloting skills. It demands a constant dance of communication, coordination, and awareness, a symphony of collaboration played out in the language of aviation. This, dear friends, is a powerful metaphor for how we interact with each other, navigating the social landscape with an array of tools and strategies.

Piloting an airplane can be a powerful metaphor for welcoming all new, incoming people into our daily lives in several ways:

- Preparation and anticipation: Like a pilot prepping an aircraft for passengers, we can prepare ourselves by setting an inviting tone, being open-minded, and learning about different cultures and experiences. This "pre-flight check" ensures a smooth landing for the newcomer.
- Navigation and guidance: Similar to a pilot helping passengers navigate turbulence, we can guide newcomers through unfamiliar situations. This involves communicating clearly, offering support, and respecting unique needs and perspectives.
- Safe landing and integration: Just as a pilot prioritizes a safe landing, we should create a welcoming environment where the newcomer feels comfortable

and accepted. This involves actively listening, showing empathy, and fostering connections with existing members.
- Shared journey and collaboration: Flying is a collaborative effort between pilots and passengers. Similarly, welcoming someone new involves collaboration. We can seek their input, share experiences, and learn from each other, enriching the lives of both parties.
- Different types of aircraft: Just as different airplanes serve different purposes, our approach to welcoming can vary depending on the context and personalities involved.
- Turbulence and challenges: Inevitably, there will be bumps in the road. Just as a pilot deals with unexpected situations, we can practice patience, understanding, and flexibility when welcoming someone new.
- Celebrating diversity: Every passenger brings unique baggage and contributes to the journey. Welcoming everyone involves celebrating this diversity and appreciating its richness.

Overall, piloting an airplane offers a dynamic and insightful metaphor for welcoming new people into our lives. By adopting a pilot's mindset of preparation, guidance, safe landing, shared journey, and adaptability, we can create a more welcoming and inclusive environment for everyone.

Things to Consider As You Encounter New People and Situations

Diversity in Age: A Strategic Asset

One of the primary opportunities presented by intergenerational teams is the diversity they bring. We will explore this topic more robustly in Chapter 5. Suffice it to note here that each generation has distinct characteristics shaped by the historical, social, and technological context of their formative years. Baby boomers bring a wealth of experience and institutional knowledge, Generation Xers contribute a balance of pragmatism and adaptability, millennials offer tech-savvy innovation, and Generation Zers bring a fresh perspective and innate digital skills.

Effective collaboration within intergenerational teams is key to leveraging these diverse skills and experiences. Bridging the communication gap between generations is crucial. While older generations may prefer traditional communication channels, younger members might be more inclined toward digital platforms. Organizations must foster an inclusive environment where all team members feel heard, valued, and respected. Encouraging mentorship programs and knowledge-sharing initiatives can facilitate the transfer of skills and insights across generations.

Technology Divide

The rapid pace of technological advancement has created a digital divide between generations. While younger team members may effortlessly navigate new technologies, older members might find it challenging. To address this, organizations should invest in training programs to upskill older team members and create a culture of continuous learning. Simultaneously, younger team members can play a pivotal role in mentoring their older counterparts, fostering a collaborative learning environment.

Balancing Leadership Styles

Intergenerational teams often grapple with differing leadership styles. Older generations might value hierarchical structures and a more formal approach while younger generations may prefer a flatter organizational hierarchy and collaborative leadership. Successful teams find a middle ground, combining the experience and wisdom of senior members with the innovative ideas and energy of younger members.

Intergenerational teams offer a rich tapestry of skills, experiences, and perspectives that can drive innovation and success in today's workforce. However, realizing the full potential of these teams requires a proactive approach to address the challenges they present. By fostering open communication, implementing mentorship programs, addressing technological gaps, and embracing diverse leadership styles, organizations can turn the complexities of intergenerational teams into a strategic advantage. In a world that thrives on diversity, these teams are the key to building resilient, adaptive, and forward-thinking organizations.

One challenge is that different generations have different communication styles and expectations. For example, baby boomers may prefer to communicate face-to-face, while Generation Z may prefer to communicate online. Bridging this gap requires a commitment to open dialogue, active listening, and a willingness to adapt communication strategies to suit diverse preferences.

Another challenge is that different generations have different work values and expectations. For example, baby boomers may value loyalty and stability while Generation Z may value flexibility and purpose.

However, intergenerational networking also presents a number of opportunities. For example, different generations can learn from each other and share their unique perspectives. Intergenerational networking can also help to break down silos and create a more inclusive workplace.

In addition, there are entire universes of demographic groups that exist without our awareness. For example, in the United States, we tend to look at anglophone personalities and networks as we go out into the world, practically ignoring the realm of Mandarin, German, Arabic, French, and Spanish speakers. Moreover, we may only focus on our known social media platforms, such as Meta, X, or Instagram, never looking at WeChat, Tencent QQ, or Baidu Tieba.

In today's dynamic landscape, the power of networking has transcended traditional boundaries and found a new home in the vast realms of social media. This shift has given rise to an intriguing phenomenon—a global stage where individuals, once anonymous, now command audiences of unprecedented scale and influence. As we delve into the intricacies of modern networking, it becomes imperative to explore the narratives of those who have mastered the art of connection in the virtual realm.

Image used under license from Shutterstock.com

Another way to consider the networks represented in the chart above, as well as your own human and digital connections—past, present, and future—is via the lens of Dr. Seuss's children's book, *Horton Hears a Who!* This book, TV show, and film franchise goes far beyond a simple tale of an elephant protecting a tiny speck of dust. Within its whimsical story lies a profound message about expanding one's network and widening one's worldview.

Horton's story is relatable. He encourages us to open up our networks and perspectives. He embraces the unseen and unheard citizens of Whoville. Horton's willingness to listen to a voice on a speck of dust, despite facing ridicule and disbelief, embodies the importance of exploring beyond the established and familiar. It encourages stepping outside one's comfort zone and seeking out diverse perspectives, even if they seem insignificant or unheard.

The Whos' world represents marginalized voices and communities often overlooked or deemed unimportant. Horton's unwavering belief in their existence and value challenges readers to acknowledge the significance of those on the periphery of their own networks. It teaches valuing diverse narratives and recognizing the potential contributions of those outside the mainstream.

Horton's efforts to protect the Whos involve building bridges and establishing communication with seemingly incompatible worlds. This symbolizes the importance of overcoming biases and preconceived notions to forge connections with diverse individuals and groups. By breaking down barriers and actively engaging with those different from ourselves, we enrich our networks and gain new perspectives.

Horton faces immense pressure and adversity in protecting the Whos. This exemplifies the challenges often faced by those advocating for marginalized communities or unheard voices. The book encourages standing up for those who may be ostracized or excluded, demonstrating the importance of using our networks to champion diversity and inclusivity.

By embracing the Whos and defying the naysayers, Horton inspires others to listen and believe. This highlights the positive ripple effect of openness and inclusivity. When we expand our networks and engage with diverse perspectives, it can inspire others to do the same, creating a more connected and understanding world.

Ultimately, *Horton Hears a Who!* is a powerful allegory for embracing diversity, challenging our preconceived notions, and actively forging connections beyond our immediate circles. By opening up our networks and worldviews, we can create a more inclusive and enriching environment for ourselves and others. By proactively engaging with different demographic groups at work or play, or by spending time on different social media platforms, we benefit from the same type of experience that Horton did.

Remember, just like Horton, who carried the Whos on a clover across the jungle, we can all play a role in championing diverse voices and expanding our own horizons by welcoming strangers into our lives, building bridges to new and unseen corners of the world.

Intergenerational Collaboration As a Networking Opportunity

Like Horton, we would be well served in our daily actions by being proactive and not waiting for people to come to us. If you reach out to people you admire and let them know that you are interested in connecting with them, chances are you will be well received. As well, when we are helpful, we are usually welcomed into someone's trust. Try more often to offer to help others with their projects or tasks. This is a great way to build relationships and demonstrate your skills. Do so consistently; don't just connect with people once and then disappear. Stay in touch with your network members on a regular basis.

Tips to Try

Here are 10 ways for different generations to better understand and interact with each other:

1. Be open-minded and respectful. It is important to listen to and learn from people of all ages, even if you do not agree with them. Try to see things from their perspective and understand why they have the beliefs and values that they do.
2. Be curious about each other's experiences. Ask questions about each other's lives, and be genuinely interested in learning about their experiences. This will help you better understand their perspectives and build rapport.

3. Find common ground. Even if you have different views on some things, there are likely to be areas where you can find common ground. Focus on these shared interests and experiences to build relationships.
4. Collaborate on shared projects. Working together on a shared goal is a great way to learn from each other and build trust. For example, different generations could collaborate on a community volunteer project or a workplace initiative.
5. Mentor each other. Mentoring is a great way to share knowledge and skills between generations. Older generations can mentor younger generations on their careers and personal development while younger generations can mentor older generations on technology and social media.
6. Spend time together socially. Socializing is a great way to get to know people better and build relationships. Invite people of all ages to social gatherings, such as parties, dinners, and outings.
7. Use technology to connect. Technology can be a great way to stay connected with people of all ages, even if you live far apart. Use social media, video chat, and other communication tools to stay in touch and share your lives with each other.
8. Learn about each other's cultures and traditions. If you are from different cultures or backgrounds, take some time to learn about each other's. This will help you better understand each other's perspectives and values.

9. Celebrate diversity. Diversity is a strength, and it is important to celebrate generational differences. This could involve attending cultural events, learning about different religions, or simply listening to the stories of people from different backgrounds. Additionally, each year commit to making new friends that have a disability, different race, creed, color, sexual orientation, or political view. Make friends with people from rival sports teams, schools, or businesses. Remember that no group is monolithic and all people are multidimensional. The adventure is in finding ways to connect with them, finding common ground.
10. Be patient and understanding. It takes time to build understanding and trust between generations. Be patient with each other, and try to see things from each other's perspectives.

Out of Your Comfort Zone

The fear of strangers is a primal echo lodged deep within us, a warning whispered down through generations. But what if it's holding us back, keeping us locked in a self-built cage of isolation, blind to the treasure trove of connections waiting just beyond?

Yes, there's an inherent risk in opening ourselves to the unknown. But consider this: every stranger you encounter was once a baby swaddled in love, eyes wide with wonder at the world. Others hold stories, experiences, and perspectives as unique and complex as yours. They are, quite simply, fellow travelers on this shared journey of existence.

So why do we let fear cloud our vision? Often, it's the unknown that spooks us. We build narratives in our heads, filling the blank pages with monsters and villains. But what if, instead of fictional demons, we imagined possibilities? Maybe that quiet guy at the café is a budding poet, your future collaborator. Perhaps the woman walking her dog has a passion for history that mirrors your own. We rarely know until we look beyond the label of "stranger," until we bridge the gap with a simple "hello."

Finding common ground, the bedrock of connection, starts with a shared humanity. We all breathe, laugh, and cry. We all yearn for love, understanding, and a sense of belonging. These are the universal threads that weave us together, no matter our cultural tapestry or life experiences. So instead of searching for differences, focus on the fundamental echoes that resonate within us all.

Of course, opening your heart takes vulnerability, exposing yourself to potential rejection, disappointment, and even hurt. But remember, these are risks we take in every relationship. And while the sting of a missed connection can linger, so can the warmth of a genuine encounter, the spark of a newfound friendship forged in the fire of shared humanity.

So the next time a stranger catches your eye, instead of flinching, smile. Ask a question, offer a helping hand, or simply acknowledge their presence. You might just unlock a hidden door, leading to a conversation that enriches your world, expands your horizons, and reminds you that in this vast mosaic of humanity, we are not so different after all.

Remember, the most enriching experiences often lie beyond the familiar, tucked away in the corners of uncharted conversations. Take a deep breath, embrace the vulnerability, and step into the exhilarating world of open hearts and unexpected connections. You might just be surprised by the treasures you find amidst the strangers.

Let your curiosity become your compass, your smile your passport, and your open heart your guide. Happy explorations!

How Do You Deal with Unexpected Encounters?

The strategic approach to networking is not simply about what we go after or aim to achieve; it's also about how we receive and process what the universe sends us. This is underscored by the lives touched and the impact created by two strangers' accidental texting.

In the vast, flickering universe of Steven Spielberg's *Close Encounters of the Third Kind,* Roy Neary's obsession with extraterrestrial signals isn't just about chasing UFOs—it's a master class in building connections with the seemingly unknowable. And as viewers, we can learn from his journey, lessons that translate beautifully into our own endeavors, both professional and personal, when navigating the uncharted territory of meeting strangers.

Roy's initial contact with the alien force is, like most first encounters, awkward and filled with misunderstanding. His obsession isolates him, leading to broken relationships and ridicule. Yet, he persists, driven by an uncanny intuition that transcends logic. This mirrors the initial discomfort we often feel when reaching out to new people—whether at a networking event, approaching a potential mentor, or striking up a conversation with a stranger on the bus. We fear misjudgment, awkwardness, and even rejection. But just like Roy, we must push past our hesitations and trust the pull of a deeper connection.

One of Roy's key strengths is his willingness to build a bridge of understanding. He studies patterns in the alien signals, learns their language, and even constructs a model of their mothership based on intuition. This dedication to understanding the "other"—whether it's a potential client from a different cultural background, a colleague with diverse experiences, or simply someone with a different worldview—is crucial for forging meaningful connections. It requires active listening, open-mindedness, and a genuine desire to see things from the other's perspective.

Roy's journey is also a testament to the power of shared experiences. His makeshift community of the obsessed, drawn together by the same unexplainable phenomenon, becomes a support system, fueling his determination and validating his experiences. This echoes the importance of finding common ground with strangers, whether through shared interests, hobbies, or even just the simple human experience of wonder and curiosity. Finding that common thread, that spark of shared passion, can build bridges even across the widest gulfs of difference.

Ultimately, *Close Encounters* reminds us that networking and building relationships with strangers isn't about mere transactions or strategic calculations. It's about venturing beyond our comfort zones, embracing the unknown, and holding onto that inexplicable urge to connect, even when faced with skepticism or ridicule. It's about understanding, shared experiences, and a deep-seated belief in the possibility of something extraordinary—whether it's a connection with the cosmos or simply the magic of forging a bond with another human being. So the next time you find yourself facing a stranger, remember Roy Neary—reach out, listen, learn, and find that common ground, and maybe, just maybe, you'll build a connection that transcends the ordinary, a connection that might just change your world.

The Look of Success

We often talk about our goals and ambitions in life or our careers by sharing examples of what success looks like.

Someone wants to be all-time football great Tom Brady or one of tennis's Williams sisters. Another person wants to be the next Miley Cyrus or Lil Nas X. It's helpful to have a sense of what success looks like at the outset of an effort. Then one can chart a more efficient way forward.

By following the advice in this chapter, we can create a more understanding and inclusive society for all generations and appreciate all the people that the universe puts in our paths. Here are some other celebrity examples of successful intergenerational collaborations:

- Beyoncé and Jay-Z have collaborated with many younger artists, including Kendrick Lamar, Rihanna, and Nicki Minaj. For example, Beyoncé and Jay-Z's 2016 album *Lemonade* featured collaborations with Kendrick Lamar and The Weeknd.
- Tony Bennett collaborated with many younger artists, including Lady Gaga, Amy Winehouse, and John Mayer. In 2014, Bennett and Lady Gaga released an album of jazz standards called *Cheek to Cheek*.
- Barbra Streisand has joined forces with many younger artists, including Josh Groban, Andrea Bocelli, and Lionel Richie. In 2016, Streisand and Groban released an album of duets called *Partn*.
- Stevie Wonder has collaborated with many younger artists, including Will.i.am, Pharrell Williams, and Ariana Grande. In 2022, Wonder's "Can't Put It in the Hands of Fate" featured Ariana Grande, John Legend, and Chris Martin.

- Elton John has teamed up with many younger artists, including Ed Sheeran, Dua Lipa, and Olly Alexander. In 2021, John released an album called *The Lockdown Sessions*, which featured collaborations with many different artists, including Sheeran, Lipa, and Alexander.
- Taylor Swift (born in 1989) has a legacy that is a shimmering constellation, each facet reflecting a different aspect of her artistic brilliance and cultural impact. Unlike many celebrities who cultivate a carefully curated image of exclusivity, Taylor embraces genuine connection. She fosters a fiercely loyal community of fans, known as "Swifties," through social media interactions, handwritten letters, and surprise meet-and-greets. She champions other artists, collaborates across genres, and uses her platform to amplify important causes. The Eras Tour is a celebration of Swift's entire career, spanning six albums and four eras. She has invited special guests from each era to join her on stage, including Haim, Phoebe Bridgers, and Maren Morris. These collaborations are a way for Swift to show her appreciation for the artists who have influenced her and also bridge the generation gap between her fans.

These are just a few examples of celebrity intergenerational collaborations. There are many other examples, and the trend is only growing. As more and more celebrities collaborate with artists from different generations, it is helping to break down stereotypes and create a more inclusive entertainment industry.

Connecting with others is an important part of life. It can make us happier, healthier, and more successful. By following the tips above, you can learn to connect with others in a meaningful way.

Discussion Questions

Consider these questions for Chapter 3:

- What are your challenges and opportunities of networking across generations?
- As you consider Taylor Swift's network, who else comes to mind as a great intergenerational networker?
- Can you identify three people in different areas of your life with whom you could grow more meaningful relationships?

Chapter 4: Matching: Connecting More Strategically

Now that we have looked at mixing up our network of relationships, let's consider how to match our relationships more strategically. It all begins with defining your purpose—your life's destination. With your purpose defined, you can strategically map your past, present, and future relationships to your own goals, ambitions, and fulfillment in life. Purpose plus strategic networking results in a life well lived and the legacy that you envision for yourself.

Our own relationship is a great example of how it works. Samir loves aviation so much that he flies and instructs others how to fly. Aviation has been central to his life and purpose. As for Toby, his stated purpose in life is making a difference through daily discovery and adventure – including piloting his own plane across the United States. Our shared love of aviation and adventure brought us together when aviation broker and friend Barron Thomas connected us in 2007.

Toby and his husband Harlan were pilots with VFR (Visual Flight Rules) ratings whereas Samir was IFR (Instrumented Flight Rules), allowing him to fly legally through clouds and inclement weather. So, Samir was Toby and Harlan's safety pilot when they flew their newly purchased Cessna from California to New York. And they have been in each others' lives ever since, including sharing family celebrations and vacations together. Aviation was the original connection point, but we invested more time and energy into the relationship for additional reasons such as adventure travel, career growth, technological innovations, and this book.

How people build meaningful networks and how they maintain meaningful networks is core to our lives and this book. We do not see popularity as the aim of skilled networking. The world is filled with popular people that have large networks by way of their ability to sing, sell, campaign or other capture peoples' attention. But few focus on people that are just great at networking.

Building and maintaining meaningful networks is about fostering genuine relationships and providing value through consistent, thoughtful interactions.

Adam Grant, an organizational psychologist, stresses the importance of authenticity, suggesting that genuine and transparent interactions build mutual trust and respect. This contrasts with simply trying to impress others, which can come off as insincere and counterproductive. Authentic connections are the foundation of a strong network.

Keith Ferrazzi, author of "Never Eat Alone," advocates for a generous approach to networking, emphasizing the importance of helping others without expecting anything in return. This philosophy, focused on giving rather than taking, naturally leads to stronger, more meaningful relationships. He emphasizes the significance of showing genuine interest in others. By asking questions about their interests and actively listening, you build rapport and trust, essential components of a solid network.

Finding common ground is another crucial element of effective networking. Richard Branson, founder of Virgin Group, advises that shared interests and passions can form the basis of enduring relationships.

Consistency in communication is also key, as highlighted by Herminia Ibarra of INSEAD. Regular, thoughtful interactions help keep relationships alive and prevent them from fading over time. Meanwhile, Reid Hoffman, co-founder of LinkedIn, suggests always thinking about how you can add value to others, whether through sharing insights, making introductions, or providing resources.

The importance of follow-up in networking cannot be overstated. Ivan Misner, founder of BNI (Business Network International), emphasizes that a timely and thoughtful follow-up message after meeting someone new can cement the connection. Gary Vaynerchuk, an entrepreneur and social media expert, recommends using social media wisely to stay connected with your network. Regularly engaging with their content and sharing valuable information helps maintain visibility and relevance.

Building deep, meaningful relationships is preferable to having a large number of superficial ones, as Sheryl Sandberg, former COO of Facebook, advises. This quality-over-quantity approach ensures that your connections are valuable and enduring. Stephen Covey, author of "The 7 Habits of Highly Effective People," supports this by advocating for win-win relationships where both parties benefit. Angela Duckworth, author of "Grit," adds that patience and persistence are vital, as building a meaningful network takes time and consistent effort.

Lastly, nurturing existing relationships is just as important as forming new ones. Brian Tracy, a motivational speaker and author, advises regularly checking in, sending notes of appreciation, or sharing relevant information to keep connections strong. Oprah Winfrey emphasizes the value of learning from mentors, suggesting that guidance from more experienced individuals can significantly enhance your professional growth. By incorporating these tips from various experts, you can build and maintain a network that is both personally and professionally rewarding.

All of these recommendations are doable by anyone. While some people may be more extroverted or more creative in their ways of engaging others, the truth is we all get better with practice. We encourage readers to try several of these approaches to networking. Just bringing your approach to how you interact to a conscious level is already going to set you up for having more meaningful, intentional relationships.

Why Purpose Matters

Finding and pursuing your purpose can bring significant benefits to your life in various ways:

- Increased well-being and happiness: Having a purpose provides a sense of direction, meaning, and fulfillment, leading to greater life satisfaction and happiness.
- Motivation and resilience: A clear purpose fuels your passion and drive, helping you overcome challenges and setbacks with greater tenacity.
- Stronger relationships: When you live with purpose, your connections with others often deepen as you connect with people who share your values and goals.
- Personal growth: The pursuit of your purpose often involves continuous learning and self-discovery, leading to personal growth and development.
- Positive impact: Aligning your purpose with helping others or contributing to a larger cause can lead to a more meaningful and impactful life.

Defining Your Purpose

Finding your purpose isn't always a straightforward journey. It requires introspection, exploration, and sometimes even trial and error. Here are some tips to help you with your search:

- Self-reflection: Ask yourself questions like "What are my values?", "What are my passions?", "What skills and talents do I possess?", and "What impact do I want to make?". Journaling and meditation can be helpful tools for introspection.
- Explore different experiences: Try new things, volunteer, travel, and engage in activities that spark your curiosity. Stepping outside your comfort zone can open doors to unexpected discoveries.
- Seek inspiration: Read biographies, listen to podcasts, or connect with inspirational people. Seeing how others live their purpose can offer valuable insights.
- Pay attention to your "flow" state: What activities make you lose track of time and bring you pure joy? These might be clues to your deeper purpose.
- Don't be afraid to experiment: Your purpose may evolve over time, so be open to new possibilities and embrace a growth mindset.
- Seek support: Talk to friends, mentors, or a therapist who can offer guidance and a sounding board for your self-discovery journey.

Remember, defining your purpose is a continuous process. Be patient and kind to yourself; and celebrate the small victories along the way. The journey itself is often as valuable as the destination.

Think of your purpose as your destination and Imagine your life as a well-mapped flight plan toward that destination where new connections represent exciting stopovers. Just like a pilot conducting a pre-flight check, take some time to define your goals. Are you seeking friendly stopovers, long-term travel buddies, or professional copilots? Knowing your purpose helps you select the right passengers for your journey.

Throughout this book, we discuss digital and analog tools that can help you network more strategically and efficiently, identifying people to meet with whom to share your joys and experiences while always advancing toward your purpose or final destination.

Regardless of the means of networking more strategically, this is why it matters: Purpose-driven networking means you have a defined purpose that acts as a powerful magnet, attracting and guiding you toward connections that propel you forward. Just as a compass directs a ship, your purpose helps you navigate the vast sea of relationships, identifying those who support your journey. Here's how:

- Alignment breeds connection: When you know your "why," it becomes effortless to identify those who share your values and aspirations. You gravitate toward individuals and groups who resonate with your purpose, creating a natural foundation for meaningful connections. These connections go beyond superficial interactions;

they foster collaboration, mentorship, and support, forming a powerful network propelling you toward your goals.
- Shared vision, shared journey: Imagine yourself and your purpose as a grand project. You can't achieve it alone. With a defined purpose, you can attract people who see your vision's value and are eager to contribute their unique skills and perspectives. This creates a synergistic effect whereby each connection strengthens the collective effort, accelerating the progress toward your legacy.
- Strategic networking with a purpose: Gone are the days of random networking events. With a defined purpose, you can target your networking efforts effectively. Seek out individuals and groups aligned with your goals, attend relevant industry events, and leverage online communities focused on your area of impact. This strategic approach ensures you're connecting with the right people, maximizing your time and potential for meaningful collaboration.
- Attracting mentors and allies: A strong purpose acts as a beacon for potential mentors and allies. Recognizing your dedication and passion, individuals with experience and expertise are more likely to offer guidance and support. These mentors can provide invaluable insights, open doors to new opportunities, and act as sounding boards on your journey.

- Building your legacy, brick by brick: Every meaningful connection you forge with a purpose in mind becomes a building block for your legacy. As you collaborate, learn, and grow alongside others who share your vision, you create a ripple effect of positive impact. This collective effort leaves a lasting mark, ensuring your purpose extends beyond yourself and inspires future generations.

Remember, the connections you make based on your purpose are not transactions; they are partnerships. Cultivate genuine relationships, offer your skills and support in return, and celebrate each other's successes. By fostering a spirit of collaboration and mutual growth, you create a network that becomes not just a means to an end but a vibrant community that shapes your legacy alongside you. So, chart your course with purpose, connect authentically, and watch as your network transforms into the wind beneath your wings, propelling you toward a fulfilling life and a meaningful legacy.

Pilots prioritize tasks and manage resources expertly. Apply this to your professional and social life. Invest your time and energy in those who truly enrich your journey, the passengers who share your flight plan's excitement. Don't hesitate to say "no" to commitments that might overload your engines. You're not obligated to be everyone's personal hot air balloon—focus on quality connections over quantity.

Like pilots constantly updating their skills, embrace lifelong learning in your professional and social life. Step outside your comfort zone, explore new social terrains, and welcome diverse perspectives. Reflect on your interactions, refine your "flight plan" as you move toward your destination, and remember, you're the captain of your own journey. By applying these pilot-inspired principles, you can navigate the world of connections with confidence, clarity, and always, with a smile that would make any air traffic controller proud.

Embark on the wondrous adventure of welcoming new people into your life. Remember, the skies are full of potential connections, and with the right mindset, you'll find yourself surrounded by a crew that enriches every step of your journey.

Helpful Resources for Planning Your Journey and Connecting the Dots

Dale Carnegie

Connecting with people is an essential skill for both personal and professional success. It allows us to build relationships, learn from others, and create opportunities for ourselves. But how do you connect with someone, especially if you don't know them very well?

Dale Carnegie famously said, "You can gain more friends in 2 months showing interest in other people than you can in 2 years trying to get other people interested in you."

Connecting with people takes time and effort, but it's one of the most important things you can do for your personal and professional life. By following these tips, you can learn to connect with anyone, regardless of their background or interests.

Dale Carnegie's *How to Win Friends and Influence People* is a useful tool for young people in the workplace to connect and build their networks in several ways.

First, Carnegie's principles can help young people make a good first impression on others and build rapport quickly. For example, he emphasizes the importance of being genuinely interested in other people, remembering their names, and making them feel important. These are all essential skills for effective networking.

Second, Carnegie's advice can help young people navigate difficult conversations and resolve conflicts peacefully. He teaches readers how to listen actively, express themselves clearly and respectfully, and find common ground. These skills are essential for building strong relationships with colleagues, supervisors, and other professionals.

Third, Carnegie's book provides guidance on how to build and maintain positive relationships with people from all walks of life. He teaches readers how to be likable, trustworthy, and influential. These are all important qualities for anyone who wants to build a successful network.

Here are some specific ways that young people can use Carnegie's principles to connect and build their workplace networks:

- Attend industry events and introduce yourself to new people. Be sure to ask questions about their work and interests and follow up after the event to stay in touch.
- Join professional organizations and get involved in activities and committees. These are great ways to meet people and build relationships with other professionals in your field.
- Reach out to people you admire and ask to connect with them. Most people are happy to help others, especially if they feel they can provide value.
- Help and support others. Offer to help with projects, share your knowledge, and connect people with needed resources.
- Be a good listener, and be genuinely interested in other people. Ask questions about their work, interests, and lives. Listen attentively to their answers, and show that you care about what they have to say.

By following Carnegie's principles, we can build strong relationships and networks that will help us achieve our career and life goals.

Imagine, for example, that Sarah, a young professional, is attending an industry event. She meets a woman named Mary who is a senior manager at a company that Sarah is interested in working for. Sarah could use the following Carnegie principles to build a relationship with Mary:

- Be genuinely interested in Mary. Sarah could ask Mary questions about her work, career path, and interests. She could also listen attentively to Mary's answers and show that she is interested in what she has to say.
- Make Mary feel important. Sarah could compliment Mary on her accomplishments and her expertise. She could also show her respect by asking for advice and feedback.
- Be helpful and supportive to Mary. Sarah could offer to help Mary with any projects or tasks on which she is working. She could also connect Mary with the resources that she needs.

By regularly practicing these Carnegie principles, Sarah could build a strong relationship with Mary. This relationship could eventually lead to a job opportunity at Mary's company or to other valuable connections in the industry.

Over 30 million copies of Carnegie's *How to Win Friends and Influence People* have been sold worldwide, making it one of the best-selling books of all time. Since 1912, it has remained a valuable resource for young people who want to connect and build their workplace networks. By following Carnegie's principles, we all can develop strong relationships with others and create networks that will help us achieve our career and personal goals.

Dr. Ruth Gotian's Networking Advice

Another great resource for connecting more strategically comes from Dr. Ruth Gotian, a psychologist, speaker, and author who specializes in introversion and networking. She is the author of the book *The Introverted Leader: Building on Your Quiet Strength* and the founder of the blog Introvert, Scientist, Mom.

Gotian has a wealth of experience with networking, and she shares her insights in her book and blog posts. Here are some of her best thoughts on networking:

- Networking is not about selling yourself. It's about building relationships. When you focus on building relationships, the rest will follow.
- Networking is not a one-way street. It's a mutually beneficial exchange. When you help others, you're more likely to get help in return.

- You don't have to be a social butterfly to be successful at networking. Even if you're shy or introverted, there are ways to network effectively.

Here are some specific tips from Dr. Gotian on how to network effectively:

- Start with small talk. Small talk is a great way to break the ice and start building a connection. When you're making small talk, ask open-ended questions, and listen carefully to the answers.
- Find common ground. Once you've gotten to know someone a little bit, look for common ground. This could be anything from shared interests to a similar background. Finding common ground will help you build a stronger connection.
- Don't be afraid to follow up. After you've met someone at a networking event, send them a follow-up email or LinkedIn message. Thank them for their time and express your interest in staying in touch.

Dr. Gotian has also written about the importance of networking for introverts. She believes that introverts can be successful at networking, but they need to approach it in a way that works for them. For example, introverts may prefer to network in small groups or one-on-one settings. They may also want to focus on building relationships with people who share their interests.

Here are some specific tips from Dr. Gotian on how introverts can network effectively:

- Set realistic goals. Do not expect to meet a hundred new people at a networking event. Set a goal to meet five or ten new people, and focus on building relationships with those people.
- Take breaks. Networking can be draining for introverts, so it's important to take breaks throughout the event. Find a quiet place to sit and recharge for a few minutes.
- Focus on building relationships. Do not worry about selling yourself or your business. Just focus on getting to know people and building relationships.

Dr. Gotian's networking advice is helpful for everyone, but it is especially valuable for introverts. By following her tips, introverts can overcome their shyness and build a strong network of professional contacts.

Key Takeaways for Human Interaction

- Tailored experiences: Every generation has unique preferences and habits. Crafting personalized experiences that cater to their behaviors and expectations is crucial.
- Authenticity is king: Across all generations, authenticity remains a cornerstone. Sincerity, transparency, and relatability build trust and loyalty.

- Multi-channel engagement: Different generations engage through various platforms and media. A multi-channel approach ensures your message reaches the right audience through their preferred channels.
- Values alignment: Understanding each generation's values and incorporating them into your brand messaging can forge powerful connections.
- Flexibility and adaptability: The generational landscape is ever-evolving. Adapting your strategies to stay relevant as new generations emerge is essential.
- Embrace technology: While some generations are more digitally savvy than others, technology impacts everyone. Incorporating tech-savvy elements can enhance engagement across the board.
- Bridge the gaps: Your campaigns should bridge generational gaps and promote unity. For instance, campaigns targeting Generation Alpha might resonate with millennial parents, creating a familial bond.

Understanding generational elements is a cornerstone of successful marketing. Acknowledging the diverse attitudes, behaviors, and preferences of baby boomers, Generation Xers, Generation Yers, millennials, Generation Zers, and Generation Alpha provides a roadmap to crafting strategies that genuinely connect. From catering to digital natives to engaging the wise, each generation offers opportunities to showcase their brand's authenticity and value. By embracing these insights and networking more strategically, marketers can navigate generational landscapes with finesse, creating lasting bonds with consumers across the spectrum.

Tower and Ground Control: Personality Inventories in the Workplace

Imagine the modern workplace as a bustling airport teeming with flights to every corner of the professional world. At the controls of each aircraft are individuals from diverse generations, each with unique flight plans and piloting styles. The baby boomers are seasoned captains, navigating by the stars of experience and precision. Gen Xers are meticulous copilots, ensuring every engine hums efficiently. Millennials are tech-savvy navigators, plotting innovative routes with their finger on the pulse of progress. And Gen Z, the fresh cadets, bring a thirst for purpose and a fresh perspective to the skies.

But just like coordinating air traffic in a busy hub, managing this multigenerational workforce can be a delicate dance. Enter the personality inventories: radar screens that reveal each pilot's strengths, blind spots, and preferred communication frequencies. Armed with this information, corporate leaders become master air traffic controllers, guiding their diverse fleets toward a smooth and collaborative flight.

Think of John, the Gen X pragmatist. His inventory reveals he thrives on clear instructions and direct communication. Maya, the millennial idealist, shines when given purpose-driven goals and flexible schedules. By understanding these nuances, leaders can pair John's efficiency with

Maya's spark, creating a team that's both grounded and visionary.

Communication becomes smoother too. No more missed landings due to cryptic emails! Understanding that John prefers face-to-face meetings and Sarah, the Gen Z whiz kid, lives on Slack, leaders can ensure everyone is on the same frequency, avoiding frustrating delays and missed connections.

But inventories are not about pigeonholing pilots into pre-defined flight paths. They're about recognizing hidden potential, like Tim, the quiet Gen Xer with a knack for data analysis, or Sarah's bubbling enthusiasm that masks untapped leadership skills. By understanding individual strengths, leaders can empower their employees to take the controls and soar to new heights.

Of course, building a harmonious airport requires more than just radar screens. Leaders must foster an inclusive environment where every pilot feels safe to be themselves. This means openly discussing generational stereotypes, addressing ageism head-on, and developing communication strategies that resonate across all age groups.

Just as a skilled air traffic controller navigates unforeseen turbulence and adapts to changing weather patterns, successful corporations must continuously refine their approach to personality inventories. Regular feedback loops and a commitment to learning and adapting ensure the skies remain clear for all generations of pilots.

Sample Inventories

Personality inventories like Myers-Briggs and the Color Inventory can be valuable tools for fostering better teamwork and individual authenticity in the workplace, but their effectiveness depends on careful application. By understanding the strengths and preferences of different personality types, teams can leverage those differences for optimal collaboration.

Introvert	**Extrovert**
"Let's Do It Right!"	"Let's Do It Now!"
* Logical	* Determined
* Organized	* Demanding
* Analytical	* Competitive
* Questioning	* Strong-willed
* Cautious	* Drive
"Let's Do It Harmoniously!"	"Let's Do It Together!"
* Relaxed	* Motivated
* Caring	* Enthusiastic
* Encouraging	* Sociable
* Patient	* Dynamic
* Sharing	* Inspire

Toby successfully introduced the Color Inventory when he joined the celebrated auction house Christie's in 2006. He organized an offsite with his communications team with the Colors Inventory as a core part of the retreat. Not only did it generate great results, but to this day, team members reference their colors and remain friends although they have all moved to other employers and countries.

The Color Inventory, a lesser-known personality assessment tool compared to Myers-Briggs, uses a unique approach based on color preferences to uncover hidden personality traits. It delves deeper than just outward behaviors, tapping into inner motivations, values, and emotional landscape. Here's how this vibrant tool can benefit different types of people in the workplace:

For the Red Personality

- Embrace your natural leadership: Reds are known for their drive, determination, and passion. The Color Inventory can help you channel these qualities into effective leadership, understanding how to motivate and inspire others while staying true to your assertive nature.
- Build stronger relationships: Reds can sometimes come across as intense or domineering. The inventory can help you identify areas where you might need to soften your approach, fostering deeper connections and understanding with colleagues.

For the Blue Personality

- Boost your confidence: Blues are often analytical and detail-oriented but can struggle with self-doubt. The Color Inventory can help you recognize your strengths and value your unique contributions, building a stronger sense of self-belief in the workplace.
- Communicate your ideas effectively: Blues can sometimes get lost in the details, making it difficult to communicate their ideas concisely. The inventory can help you develop strategies for clear and impactful communication, ensuring your valuable insights are heard.

For the Green Personality

- Find your creative voice: Greens are naturally artistic and empathetic but can struggle to express themselves in the corporate world. The Color Inventory can help identify your unique creative strengths and find ways to integrate them into your work, fostering a more fulfilling and authentic work experience.
- Navigate conflict constructively: Greens can be peacekeepers but sometimes struggle to assert themselves in conflict situations. The inventory can equip you with tools for navigating disagreements in a collaborative and productive way.

For the Yellow Personality

- Channel your optimism: Yellows are sunshine personified, but their constant cheerfulness can sometimes be misconstrued. The Color Inventory can assist you in finding the right balance, expressing your positive energy while maintaining professionalism and respecting boundaries.
- Stay grounded: Yellows can get caught up in their enthusiasm, overlooking important details. The inventory can help you develop a more grounded approach, ensuring you deliver quality work while maintaining your infectious optimism.

Remember, the Color Inventory is just one tool in the vast toolbox of self-discovery. Its true power lies in its ability to spark conversations, encourage self-reflection, and ultimately, empower individuals to bring their authentic selves to work, fostering a more inclusive and productive environment. So, embrace the colors within you, and let them guide you toward a more fulfilling and meaningful work experience!

Image used under license from Shutterstock.com

Myers-Briggs in a Nutshell: Your Personality Cockpit

Imagine your personality as a sleek, personalized aircraft navigating the skies of human interaction. The Myers-Briggs Type Indicator (MBTI) is your cockpit dashboard, revealing four key dials that shape your flight:

- Introversion/Extraversion: This gauges your energy source. Are you a self-recharging loner

(Introvert) or a social butterfly fueled by external interaction (Extravert)?
- Sensing/Intuition: This reflects how you process information. Do you prefer concrete facts and details (Sensing) or abstract ideas and possibilities (Intuition)?
- Thinking/Feeling: This shows how you make decisions. Do you prioritize logic and objectivity (Thinking) or personal values and empathy (Feeling)?
- Judging/Perceiving: This reveals your approach to life. Do you like structure and planning (Judging) or prefer flexibility and spontaneity (Perceiving)?

By combining these four dials, the MBTI identifies 16 distinct personality types, each with unique strengths, weaknesses, and preferences. It's not about labeling you, but rather providing a framework to understand yourself and others better.

The MBTI is not a rigid system. Your dials can change over time and in different situations. It's a tool for self-exploration, not a definitive destination. The MBTI can be a valuable guide in understanding your strengths, improving communication, and building stronger connections with the diverse passengers on your life journey.

The self-awareness from any personality inventory allows individuals to play to their natural strengths and avoid feeling pigeonholed into roles that don't suit them, ultimately leading to greater individual satisfaction and team productivity.

However, it's crucial to remember that personality inventories are not static boxes. They provide a framework for understanding, but individual personalities are nuanced and multifaceted. Reducing people to mere labels can be counterproductive, stifling creativity and discouraging individuals from stepping outside their perceived "type."

Instead, use these tools as conversation starters, encouraging open communication and appreciation for individual differences. When used thoughtfully, personality inventories can help people navigate the complexities of teamwork, build bridges between diverse perspectives, and eventually create a work environment where everyone feels empowered to be their authentic selves.

Ultimately, by embracing personality inventories as tools for understanding and collaboration, corporations can transform their diverse workforce into a high-performing fleet, each aircraft contributing its unique strengths to achieve a shared destination: a thriving, harmonious, and truly intergenerational workplace.

So open the communication channels, trust your instruments, and prepare for a smooth and collaborative journey across the skies of professional success.

Remember, with the right approach, every generation can be a top gun in the team effort to reach new heights.

From the Abstract to the Concrete: Operating at a Higher Level

As we delve into personality types like those described in the Myers-Briggs indicator and the Color inventory, we gain valuable insights into human behavior; however, understanding personalities is just the beginning. To forge deeper connections, we must actively seek common ground.

Most introductions are transactional. We meet someone, learn their name, and make a quick mental note of if and how we might want to know them better. Maybe they're a potential new friend, customer, employer, or investor. Maybe they're just someone we are intrigued by.

The richest, most memorable introductions happen when we go beyond the perfunctory or transactional and find more in common with the other person. This could be anything from shared interests and hobbies to similar experiences and values.

One effective strategy is finding three or more points of connection with individuals. This goes beyond small talk, focusing on shared interests or experiences. By actively listening and uncovering connections, we lay the groundwork for more meaningful interactions.

In the "rule of three," Toby advises finding three points of connection with an individual when meeting them. It will likely start with the matter at hand—a transaction, social interaction, or random meeting. The connection might be made stronger with a second point of connection, such as a common bond over where both people went to school or a shared love of a restaurant, sports team, or dog breed. Identifying a third point of connection completes a triangle that makes you more memorable and likable.

When you introduce yourself or you introduce two other people with three points in common, you're setting everyone up for success. You're creating a foundation to build on, and you're making it more likely that a meaningful conversation and relationship will result.

So, the next time you're meeting someone new or introducing two people, take the time to find three points in common. It's a small investment that will ultimately pay off. Finding three points in common takes a little effort, but it's worth it. When you take the time to get to know someone on a deeper level, you build trust and rapport. You also make it more likely that you'll stay in touch and develop a meaningful relationship.

Here are a few examples of how to introduce two people with three points in common:

- Two people who are interested in sustainable living: "Sarah, I'd like to introduce you to my friend John. He's also passionate about sustainable living. John, Sarah is

a vegan chef, and she's starting her own business selling organic produce. I know you two will have a lot to talk about."
- Two people who are new to the city: "Mike, I'd like to introduce you to my friend Mary. She's also new to the city. Mary, Mike is a software engineer, and he's looking for a new hiking buddy. I know you two have a lot in common."
- Two people who are parents of young children: "Jessica, I'd like to introduce you to my friend Dave. He's also the father of a two-year-old. Dave, Jessica is a stay-at-home mom, and she's looking for a new playgroup for her son. I know you two will have a lot to discuss."

Of course, as the person introducing two individuals over shared points of interest, you are the common denominator, and thereby a third point of connection.

Similarly, when you meet someone for the first time, try to discern a few points of mutual interest. For example:

- Pay attention to their body language and nonverbal cues. What are they wearing? What kind of jewelry do they have? What are their facial expressions like? What are their interests?
- Ask open-ended questions. Avoid yes/no questions and focus on questions that require the other person to elaborate. For example, instead of asking "Do you like

to travel?", ask "What's your favorite place you've traveled to?"
- Listen carefully to their answers. Do not just wait for your turn to talk. Pay attention to what they're saying and ask follow-up questions.
- Look for common ground. It could be anything from a shared interest in sports or music to a similar experience growing up.
- Don't be afraid to be vulnerable. Sharing something personal about yourself can help you connect with the other person on a deeper level.

Here are some sample open-ended questions that you can use to start a conversation. What brings you here today?

- What are you most looking forward to at this event?
- What brings you here today?
- What's your favorite thing about working in this industry?
- What are some of your hobbies?
- What are you passionate about?
- What's the most interesting thing you've learned recently?
- If you could travel anywhere in the world, where would you go?
- What's your favorite book/movie/TV show?
- What's your favorite food?
- If you could have any superpower, what would it be?

- What's the most important lesson you've learned in life?

The Power of Listening: Unlocking Understanding with Active Listening

In a world filled with noise and distractions, truly listening can feel like a lost art. But in the realm of communication, few skills hold the power to transform relationships and build bridges like active listening. It's more than just hearing the words; it's about immersing yourself in the speaker's world, seeking to understand not just what they say but also the emotions and meaning woven into their message.

Active listening is a deliberate act of focus. It's putting aside your own agenda, silencing the internal chatter, and tuning your senses to the speaker. It's about paying attention not just to the words but also to the nonverbal cues—the facial expressions, the gestures, and the tone of voice. These can often tell a different story than the spoken words themselves.

The benefits of active listening are manifold, particularly as you build your networks. It fosters empathy and connection, allowing you to truly see and understand the other person's perspective. It builds trust and strengthens relationships, showing that you value what they have to say. It can defuse conflict and navigate challenging conversations, allowing you to calmly address underlying emotions and concerns.

But how do you become an active listener? Here are some key practices:

- Be present: Put away distractions like your phone and maintain eye contact. Show genuine interest in the speaker through your body language.
- Listen with your whole self: Pay attention to the speaker's emotions, the rhythm of their speech, and any unspoken messages their body language may convey.
- Ask open-ended questions: Encourage the speaker to elaborate, rather than simply offering yes/no options. Phrases like "Can you tell me more about that?" or "How did that make you feel?" can unlock deeper understanding.
- Paraphrase and reflect: Sum up what you've heard in your own words to ensure understanding. This shows the speaker that you're paying attention and helps clarify any misunderstandings.
- Avoid interrupting or giving unsolicited advice: Let the speaker finish their thoughts before sharing your own. Offer support and understanding; but resist the urge to jump in with solutions or judgments.

Active listening is a continuous practice, not a one-time skill. The more you hone it, the more you'll discover its power to transform your relationships, both personal and professional. By opening your ears and heart to truly listen, you unlock a world of understanding, connection, and meaningful communication.

In the age of constant information overload, make a conscious choice to embrace active listening. Give others the gift of your full attention, and watch the magic of connection unfold. Remember, sometimes the most powerful words are the ones we listen to, not the ones we speak.

Once you've established three points of connectivity with a new person and you've engaged them sincerely with active listening, you are on your way to building lasting rapport.

Mastering the art of finding three points of connection and listening actively not only facilitates deeper interactions in the moment but also lays a solid foundation for maintaining relationships over time. After all, meaningful connections are not just about the initial encounter; they're about nurturing those connections into lasting relationships.

Following Through

Once you've established common ground with someone, the next step is ensuring that the connection doesn't fade away after your initial meeting. Staying in touch with people you've met requires a proactive approach and a genuine interest in cultivating relationships beyond the surface level.

In the upcoming sections, we'll explore practical strategies for effectively staying in touch with individuals with whom you've connected. From leveraging digital communication tools to scheduling regular check-ins, these tips will help you nurture your network and foster long-lasting connections.

How well you maintain your relationships will determine your reputational currency. Your reputation can rise or fall vis-a-vis others based on your actions or inaction. Beyond the initial connection with someone new, you need to consider if and how you will build on the introduction to deepen your relationship over time in a natural, unforced way. Aligning your network with your purpose in life ensures a stronger legacy.

Some common ways to stay in touch and grow your relationships include:

- Following up after the conversation: Send the other person an email or text or connect with them on social media. This shows that you're interested in staying in touch and learning more about them. If you have a photo, newsclip, or story to share from your conversation, this is the perfect way to relay it and remind the person of how you connected. One trick

that Toby enjoys is grabbing a selfie after a fun interaction and having the other(s) text or email it to themselves with his phone. Not only does it capture, document, and share a pleasant moment, but it also gives him contact information for future exchanges.
- Finding ways to stay connected: If you have common interests, see if there are any events or groups that you can attend together. If you live close to each other, invite them out for coffee or lunch.
- Making an effort: Staying in touch takes effort, but it's worth it to maintain meaningful relationships. Fortunately, digital tools such as calendar reminders and social media allow us to keep in touch over shared interests and moments in time, such as a birthday, anniversary, game, or concert.

A Word About Respect

Perhaps the most important point to make here is to always be respectful. Treat everyone with kindness and consideration, regardless of their background or social status. We all know that feeling of not being respected, seen, or heard almost more than we remember when we do feel included or respected.

Nelson Mandela was able to grow his network and impact by building relationships with people from all walks of life, including both his supporters and his opponents. He was known for his ability to listen to others and find common ground, even with those who disagreed with him. This allowed him to build a broad coalition of support for his cause, which was essential to his success.

For example, Mandela formed a close friendship with F.W. de Klerk, the white South African president who released him from prison in 1990. The two men worked together to negotiate the end of apartheid and create a new democratic South Africa. Mandela's ability to build rapport with de Klerk was essential to the success of these negotiations.

Mandela also built relationships with people from other countries, including world leaders such as Bill Clinton and Tony Blair. This helped raise awareness of the apartheid struggle and build international support for Mandela's cause.

Similarly, Mother Teresa was able to grow her network and impact by focusing on the needs of the poor and marginalized. She founded the Missionaries of Charity in 1950, which began with a small group of women who cared for the sick and dying in Kolkata, India. Over time, the Missionaries of Charity grew into a global organization with over 5,000 sisters working in over 100 countries.

Mother Teresa's focus on the poor and marginalized helped her build a network of supporters all over the world. People were drawn to her compassion and commitment to helping those in need. This network of supporters helped Mother Teresa expand her work and reach more people in need.

Martin Luther King, Jr., was able to grow his network and impact by using his voice to speak out against injustice. He was a powerful advocate for civil rights and equality. King's speeches and writings inspired millions of people around the world to join the fight for justice.

King also built relationships with other leaders in the civil rights movement, such as Rosa Parks and Ralph Abernathy. He worked with these leaders to organize protests and boycotts and to lobby the government for change. King's network of relationships was essential to the success of the civil rights movement.

The examples of Nelson Mandela, Mother Teresa, and Martin Luther King, Jr. show us that we can grow our networks and make a positive impact on the world by:

- Building relationships with people from all walks of life, even those whom we consider polar opposites of ourselves.
- Focusing on the needs of others.
- Using our voices to speak out against injustice.
- Working with others to achieve common goals.

By following these examples, we can all make a difference in the world.

Again, in strategically building your network, always practice being non-judgmental. Everyone has flaws and makes mistakes. Be forgiving, understanding, and supportive. Be there for your friends and family when they need you. And keep practicing. By law, pilots have to complete at least three successful takeoffs and landings every 90 days. Be sure to practice just as many or more engagements with strangers each quarter as you build your network and legacy.

Discussion Questions

For Chapter 4, consider these questions:

- Do you view Dale Carnegie's advice on networking to be contemporary and relevant to your life? Why?
- Consider your current network. Who came into it by chance, and who did you target? Who have you continued to invest in developing, and who has gone by the wayside? How and why?
- List 10 people that you would like to add to your personal and/or professional network this year; and suggest how you might go about making the connections.

Chapter 5: Diversity

Turbulence and Takeoff: Our Pilot's Case for Diversity

In the cockpit, just as in life, relying solely on our own perspective, our own limited experience, could be disastrous.

That's why, after years spent as solo pilots, we became fervent advocates for diversity. It's not just a buzzword for us; it's a lifeline forged in the crucible of the sky. Here's why:

- Multiple instruments, richer picture: Imagine two sets of eyes on the horizon, two minds interpreting the dance of gauges and weather charts. That's the power of diversity. Every copilot we have flown with has brought a unique perspective, a different nuance to the unfolding situation. Their voices, calm and seasoned, have pulled me back from the brink during emergencies, and their insights have steered us clear of unseen storm cells. Diversity isn't just about numbers; it's about expanding the cockpit of perception, seeing the world through a tapestry of experiences woven from different backgrounds and disciplines.
- Blind spots minimized; errors reduced: We all have blind spots—cognitive biases that can distort our judgment. But when you have a copilot with a

different frame of reference, those blind spots become illuminated. They call out your assumptions, challenge your interpretations, and together, you forge a path that's more nuanced, more likely to withstand the turbulence. Diversity isn't about pointing fingers; it's about shared vigilance, a collective effort to ensure no crucial detail gets lost in the clouds.

- Innovation takes flight: Imagine tackling a complex landing in a stale cockpit, the same routines playing out again and again. Now, picture fresh ideas bouncing off the walls, fueled by diverse experiences and perspectives. That's the magic of a team on which "different" isn't seen as a liability but as a launchpad for innovation. Diversity isn't just about inclusion; it's about unlocking the collective genius, the unexpected solutions that arise from the friction of contrasting viewpoints.

Our cockpits have been our classrooms, teaching us the invaluable lesson that to truly soar, we need to fly in formations, not solo. Embracing diversity isn't just a moral imperative; it's a survival tactic. It's the difference between weathering a storm and succumbing to it. So the next time you hear the call for diversity, remember this pilot's plea: open the cockpit doors, invite different voices in, and watch our collective journey reach new heights, together.

Diversity is important in every aspect of our lives, and networking is no exception. When we connect with people from different backgrounds, we expose ourselves to new ideas, perspectives, and opportunities. This can help us grow as individuals and achieve our professional goals. Some of the benefits of having a diverse network include:

- Increased creativity and innovation: When we work with people from different backgrounds, we bring together a wider range of ideas and perspectives. This can lead to more creative and innovative solutions to problems.
- Improved decision-making: When we have access to a variety of perspectives, we are better able to make informed decisions. This is especially important in the workplace, where we are often faced with complex challenges.
- New opportunities: A diverse network can open up new opportunities, both personally and professionally. For example, we may be introduced to new job opportunities, new clients, or new collaborators.
- Personal growth: Networking with people from different backgrounds can help us learn about different cultures and perspectives. This can help us grow as individuals and become more open-minded and tolerant.

In addition to these benefits, diversity in our networks is also important because it helps create and reflect a more civil, inclusive, and sustainable society for everyone. Diversity in networking begins with a conscious effort to expand cultural horizons. It's also one of the most thrilling adventures in life.

Embracing cultural intelligence involves recognizing the value of different cultural backgrounds and understanding how they can contribute to a richer, more diverse network.

Inside organizations, it is important to foster cultural intelligence—helping a group develop an awareness of cultural nuances and customs to foster understanding. Organizations also need to foster adaptability by helping members cultivate the ability to adapt communication styles to resonate with individuals from diverse cultural backgrounds. Organizations also need to foster open-mindedness, encouraging interactions with an open mind, acknowledging that different perspectives can offer unique insights.

More broadly, organizations should encourage learning from global perspectives—international networking with connections beyond local boundaries to engage with professionals from different countries and regions. Cross-cultural collaboration encourages collaborative projects involving individuals with diverse cultural perspectives to capitalize on a wealth of ideas.

Creating inclusive networking events is essential for fostering a diverse and welcoming environment. By intentionally designing gatherings that appeal to a broad audience, you can ensure that people from various backgrounds feel comfortable and valued.

Organizations are stronger when they encourage diversity in attendees at their events. They use targeted Invitations to reach out to professionals from underrepresented groups to ensure a diverse attendee list. They promote inclusivity and clearly communicate that the event is open to everyone, regardless of background, to encourage broad participation. They create safe spaces for open discussions of inclusive topics and facilitate activities, structures, and practices for networking, ensuring that everyone has an opportunity to engage in meaningful conversations.

By integrating DEI principles into networking efforts, individuals and organizations can build richer, more dynamic, and more effective connections. Several prominent figures have championed Diversity, Equity, and Inclusion (DEI) in their professional environments and in their own networking, demonstrating the significant impact of these principles.

Ursula Burns, former CEO of Xerox, was the first African-American woman to lead a Fortune 500 company. She implemented policies at Xerox that focused on promoting diversity and creating a more inclusive workplace. Under her leadership, Xerox was recognized for its commitment to diversity, notably increasing the representation of women and minorities in leadership positions.

Satya Nadella, CEO of Microsoft, has been a strong advocate for diversity and inclusion at Microsoft. He has focused on transforming the company's culture to be more inclusive, ensuring that diverse voices are heard and valued. Nadella implemented various initiatives, such as unconscious bias training and a more inclusive hiring process, leading to increased diversity within the company.

Apple CEO Tim Cook has been vocal about the importance of diversity and inclusion at Apple. He has led efforts to increase the representation of women and minorities within the company and has publicly supported initiatives promoting LGBTQ+ rights. Cook's leadership has made Apple a more inclusive and equitable workplace, reflecting his commitment to these values.

PepsiCo's former CEO Indra Nooyi emphasized the importance of diversity and inclusion during her tenure at PepsiCo. She implemented policies that promoted gender diversity and worked to ensure that women and minorities had opportunities for advancement. Her efforts significantly improved the company's diversity metrics and created a more inclusive corporate culture.

Marc Benioff, CEO of Salesforce, has been a strong proponent of pay equity and gender equality at Salesforce. He conducted company-wide audits to address and eliminate pay disparities between men and women. Benioff has also been an advocate for broader diversity initiatives, including increasing the representation of underrepresented groups within the company.

Ken Frazier, former CEO of Merck & Co, has been an advocate for diversity and inclusion throughout his career. At Merck, he emphasized the importance of creating an inclusive workplace and worked to ensure that the company's leadership reflected a diverse set of backgrounds and perspectives. Frazier has also been active in broader societal discussions about racial equality and justice.

These leaders have not only promoted DEI within their organizations but have also served as influential role models, demonstrating how prioritizing diversity, equity, and inclusion can lead to more innovative, successful, and equitable workplaces.

Here are some tips for building a more diverse network:

- Expand your horizons: Attend events and conferences that are outside your comfort zone. This is a great way to meet people from different backgrounds and industries.
- Be open-minded: When you meet new people, be open to learning about their experiences and perspectives. Don't be afraid to ask questions and challenge your own assumptions.
- Be inclusive: Make a conscious effort to connect with people from different backgrounds. Don't just gravitate toward people who are similar to you.
- Be supportive: Be a champion for diversity in your network. Amplify the voices of underrepresented groups and help them connect with each other.
- Leverage AI: You can privately practice a dialogue with any historical figure or person from a different demographic background on any topic with

generative AI. For example, prompt Bard.Google.com to explain to you as a white male how to discuss structural racism, or prompt Chat.GPT to describe Charles Darwin's views on veganism or explain ways Muslims and Hindus might get along.

Practical Steps to Foster DEI in Networking:

- Seek out diverse groups and events: Attend conferences, seminars, and meet-ups that focus on or include a wide range of participants.
- Mentorship and Sponsorship: Actively mentor and sponsor individuals from underrepresented groups.
- Inclusive Communication: Use inclusive language and be mindful of different communication styles.
- Active Listening: Listen actively and empathetically to understand different perspectives.
- Challenge Biases: Be aware of and challenge your own biases and encourage others to do the same.

The Multiplier Effect: Why Inclusion Beats Exclusivity

Here's another reason why embracing inclusion in your network can unlock a hidden superpower: the multiplier effect.

In the cutthroat world of professional networking, many believe that exclusivity is the key to success. They curate meticulously crafted circles, attend high-priced events, and drop names with the finesse of Olympic hopefuls. But what if the real gold lies not in guarded fortresses, but in wide-open doors? What if the true path to professional growth lies not in exclusion, but in inclusion?

Imagine your network as a web. With each connection, you add another strand, strengthening the overall structure. But with exclusion, you create gaps, weak points where opportunities can slip through. Inclusion, on the other hand, weaves a tighter, more resilient web. It connects you to diverse perspectives, unexpected talents, and hidden gems you might never have encountered in your exclusive circles.

Consider this:

- Diversity of thought: When you welcome people from different backgrounds and experiences, you tap into a wellspring of creativity and innovation. Different viewpoints challenge assumptions, spark new ideas, and lead to solutions you might never have conceived on your own.
- Unexpected connections: The person you meet at a casual coffee meetup might introduce you to the CEO you've been trying to reach for months. The intern you mentor might become your future business partner. Inclusion throws open the doors to serendipity, connecting you with people who can propel you forward in unexpected ways.
- The ripple effect: When you include others in your network, you don't just benefit yourself; you benefit everyone. Your connections gain access to your own

network, creating a domino effect of opportunity and growth. The more inclusive you are, the stronger and more vibrant the entire ecosystem becomes.

In an exclusive network, your success is limited by the size of your circle. But in an inclusive network, your success is multiplied by the success of everyone with whom you connect. The more you give, the more you receive, and the entire network thrives as a result.

Of course, inclusion doesn't mean abandoning all standards. It's about quality, not quantity. Surround yourself with people who share your values, inspire you, and challenge you to be your best. But don't let arbitrary barriers like background, affiliation, or status hold you back from connecting with those who can truly enrich your professional journey.

So the next time you're tempted to close the door on your network, remember the power of the multiplier effect. Embrace inclusion, weave a stronger web, and watch your professional horizons expand in ways you never imagined. Because in the end, it's not who you keep out that matters, but who you let in.

Building a diverse network requires intentional efforts to break down barriers, celebrate differences, and create an atmosphere where individuals from all walks of life feel welcomed and valued. By actively seeking out and embracing diversity, you enrich your network with a variety of perspectives, experiences, and opportunities for collaboration.

Discussion Questions

For Chapter 5, consider these questions for building diverse and inclusive networks:

- How has your understanding of diversity and inclusion evolved throughout your career? Has your approach to networking shifted as a result?
- Think about three individuals in your network who bring unique perspectives or experiences. How have they challenged your assumptions and contributed to your personal and professional growth?
- What practical steps can you take to actively build your network to reflect diversity, equity, and inclusion? Consider resources, platforms, and strategies you can use.
- Imagine navigating a networking event where you feel like the "other." How can you leverage your strengths and unique perspective to build meaningful connections despite potential biases?
- Share an example of a time when you actively advocated for someone from an underrepresented group in your network. What impact did your action have?

Example of Hedcut Image

Image used under license from Shutterstock.com

Building Trust Through Hedcut Images: *The Wall Street Journal's* Legacy of Visual Storytelling. For over 40 years, *The Wall Street Journal* (WSJ) has employed a unique visual approach to building trust with its readers: the intricate art of hedcut imagery. These handcrafted, stipple-style portraits and illustrations—literally connections of black and white dots—have transcended the realm of mere decoration, becoming a cornerstone of the WSJ's brand and a powerful tool for conveying complex information and fostering a sense of connection with its audience.

Chapter 6: Trust: The Gold Standard

In a world driven by fleeting interactions, shrinking attention spans, and digital connections, the importance of genuine, trusting human relationships has never been more vital. Trust serves as the bedrock upon which strong, lasting connections are built. It allows us to feel safe, secure, and understood, opening the door for vulnerability, intimacy, and growth.

However, trust is not a static state. It is a dynamic, fragile thing that requires constant cultivation and maintenance. A single act of betrayal, a careless word, or a broken promise can shatter trust in an instant, leaving behind a trail of hurt, resentment, and doubt.

The consequences of broken trust can be far-reaching, impacting not only our personal lives but also our professional endeavors and overall sense of well-being. It can breed suspicion, impede collaboration, and create a toxic environment where genuine connection feels impossible.

Fortunately, we are not powerless in this regard. By understanding the fundamental pillars of trust—empathy, logic, and authenticity—we can actively cultivate this valuable quality in our interactions with others. Specifically:

- Empathy allows us to see the world through another person's eyes, understanding their feelings and perspectives with genuine compassion.

- Logic demands consistency and reliability in our actions, ensuring that our words and deeds reflect our true intentions.
- Authenticity requires us to embrace our true selves, flaws and all, fostering genuine connection and vulnerability.

By nurturing these three essential pillars, we can build and maintain trust in our relationships, paving the way for deeper connections, meaningful collaboration, and more fulfilling lives.

Pillar 1: Empathy: Seeing the World Through Another's Eyes

Empathy, the ability to understand and share the feelings of another, forms the cornerstone of trust in human connections. It acts as a bridge across differences, allowing us to connect with others on a deeper level and build relationships based on mutual understanding and respect.

There are three primary types of empathy:

1. Cognitive empathy: Understanding the other person's mental state and perspective, even if you don't experience their emotions yourself. This involves actively listening, paying attention to their words and body language, and attempting to see things from their point of view.
2. Affective empathy: Sharing the emotional state of the other person, feeling their emotions as your own. This

goes beyond simply understanding their feelings and involves experiencing a similar emotional response.
3. Compassionate empathy: Not only understanding and sharing the other person's emotions but also feeling a desire to help them alleviate their suffering. This manifests as a willingness to offer support, comfort, and assistance.

Developing these different types of empathy is crucial for building trust. Here are some practical tips to help you cultivate this essential skill:

- Actively listening: Truly listen to what the other person is saying, focusing on their words, tone, and body language. Avoid interrupting and encourage them to elaborate by asking open-ended questions.
- Observing non-verbal cues: Pay attention to the other person's facial expressions, posture, and gestures. These can often reveal unspoken emotions and provide valuable insights into their true feelings.
- Asking open-ended questions: Instead of questions that can be answered with a simple yes or no, use open-ended questions to encourage the other person to share their thoughts, feelings, and experiences.
- Showing genuine interest: Be genuinely curious about the other person's life, experiences, and feelings. Ask follow-up questions and show that you are engaged in their conversation.
- Avoiding judgment and criticism: Create a safe space for the other person to share their feelings without fear

of judgment or criticism. Accept their perspective, even if you don't agree with everything they say.

Practice Idea: The Empathy Mirror Exercise

This exercise is a powerful tool for developing empathy in your relationships. Find a partner and take turns sharing personal experiences or emotions. As you listen to your partner, practice reflecting back their emotions and perspectives using phrases like, "It sounds like you're feeling..." or "I can understand why you would feel that way." This simple act of mirroring can deepen understanding, build trust, and strengthen your connection.

By actively engaging in these practices and cultivating empathy in your interactions with others, you can lay the foundation for strong, trusting relationships that will enrich your life in countless ways. Remember, empathy is a journey, not a destination. Be patient with yourself and others, and celebrate the small victories along the way.

Pillar 2: Logic: Building Trust Through Consistent Actions

While empathy lays the emotional groundwork for trust, logic serves as the sturdy framework that supports it. It is the consistent, predictable aspect of our behavior that allows others to rely on us and feel confident in our intentions.

Imagine a friend who promises to help you move but cancels at the last minute without a valid reason. Or a colleague who constantly misses deadlines and delivers subpar work. These experiences can create an environment of distrust, making it difficult to feel secure or confident in the relationship.

On the other hand, consistency and reliability foster trust. When we keep our promises, follow through on commitments, and demonstrate responsible behavior, we show others that they can depend on us. This builds a sense of security and predictability that allows relationships to thrive.

Here are some key strategies for building trust through logic:

- Keep promises and commitments: When you make a promise, stick to it. If something unexpected arises and you need to change your plans, communicate honestly and proactively to explain the situation and find a solution together.
- Communicate clearly and concisely: Be clear and direct in your communication, avoiding ambiguity and vagueness. This helps others understand your intentions and reduces the risk of misunderstandings.
- Be transparent and honest: Openness and transparency are crucial for building trust. Be honest about your mistakes and shortcomings; and avoid hiding information or engaging in deception.
- Take responsibility for mistakes: Everyone makes mistakes. What matters is how you handle them. Take responsibility for your actions, apologize when necessary, and make amends to rectify the situation.
- Demonstrate good judgment: When faced with difficult decisions, take the time to consider all sides of

the issue, and act with sound judgment. This shows others that you are capable of making responsible choices and can be trusted with important matters.

Unique Idea: The Logic Grid Tool

This tool can help you analyze your decision-making processes and identify areas for improvement. Create a grid with columns for factors such as potential consequences, ethical considerations, and value alignment. When faced with a decision, list each option and evaluate it based on these criteria. This can help you make more logical and consistent choices, thereby strengthening trust in your relationships.

Remember, building trust through logic is a continuous process. It requires ongoing commitment and effort to demonstrate your reliability and consistency in every interaction. By integrating these strategies into your daily life, you can cultivate a reputation as someone who is trustworthy and dependable, paving the way for deeper, more meaningful connections.

Pillar 3: Authenticity: Unveiling Your True Self

The final pillar of trust, authenticity, is the heart of genuine connection. It's about embracing your true self, flaws and all, and allowing others to see and accept you as you are. This vulnerability fosters deep trust and intimacy, creating the space for meaningful relationships to flourish.

When we wear masks and hide our authentic selves, it creates a barrier between us and others. We hesitate to share our true thoughts and feelings, fearing rejection or judgment. This lack of authenticity can lead to shallow relationships and feelings of isolation.

In contrast, when we embrace our true selves, we invite deeper connection and intimacy. By sharing our vulnerabilities, we create a space for others to do the same, fostering empathy, understanding, and trust.

Reputational Currency: The Value of Trust in an Age of Information

In today's interconnected world, reputation has become a powerful currency, influencing our interactions, opportunities, and overall success. Just as traditional currencies like the dollar or euro hold value and facilitate transactions, our reputation serves as a social currency that shapes our relationships and outcomes.

Reputational currency is the collective perception of an individual or entity's trustworthiness, competence, and integrity. It's the intangible asset that builds trust and opens doors to opportunities. A positive reputation attracts favorable interactions, collaborations, and support while a negative reputation can hinder progress and limit possibilities.

The rise of the digital age has amplified the significance of reputational currency. Online platforms, social media, and review sites have created a global marketplace for reputation, where our actions and interactions leave a permanent digital footprint. This increased visibility has made reputation management a crucial aspect of personal and professional branding.

In the realm of business, reputational currency is a key differentiator. A company with a strong reputation enjoys customer loyalty, attracts top talent, and secures favorable partnerships. Conversely, a company with a tarnished reputation faces customer skepticism, employee defections, and business setbacks.

The concept of reputational currency extends beyond individuals and businesses to encompass organizations, communities, and even nations. A country's reputation influences its ability to attract foreign investment, forge diplomatic alliances, and secure international cooperation.

Building a strong reputational currency requires consistent effort and adherence to ethical principles. It's about demonstrating honesty, integrity, and a commitment to excellence in all endeavors. It's about delivering on promises, respecting commitments, and treating others with fairness and respect.

In an increasingly interconnected world where trust is a scarce commodity, reputational currency is a valuable asset.

By cultivating a positive reputation, individuals, businesses, and organizations can enhance their standing, expand their opportunities, and achieve greater success.

Trust is the foundation of any strong relationship. It is the belief that someone is reliable, honest, and has your best interests at heart. Trust is essential for both personal and professional relationships. It allows us to feel comfortable sharing our thoughts and feelings, collaborate effectively, and achieve our goals.

Trust is important because it allows us to:

- Feel safe and vulnerable. When we trust someone, we feel safe to be ourselves and share our thoughts and feelings without fear of judgment.
- Collaborate effectively. Trust is essential for effective collaboration. When we trust our teammates, we are more likely to share ideas, work together toward common goals, and support each other.
- Achieve our goals. Trust can help us achieve our goals by providing us with the necessary support and resources. For example, if we trust our manager, we are more likely to ask for help and support when we need it.

Dr. Brené Brown defines vulnerability as "uncertainty, risk, and emotional exposure." She explains that vulnerability is not weakness, but rather courageously opening up to the world, allowing yourself to be seen and experienced. It's about being willing to be hurt, judged, criticized, and rejected.

Brown argues that vulnerability is the birthplace of love, belonging, joy, courage, empathy, and creativity. It's the foundation of any meaningful connection. When we're vulnerable, we allow ourselves to feel deeply and authentically, which opens us up to experiencing the full range of human emotions.

Brown also emphasizes that vulnerability is not optional. We all experience it, whether we choose to acknowledge it or not. The difference is whether we choose to embrace vulnerability and use it to our advantage or whether we try to avoid it and shield ourselves from it.

Here are some of the key insights from Brown's work on vulnerability:

- Vulnerability is not weakness. Vulnerability is courage. It takes courage to open yourself up to the world, be seen and experienced, and risk being hurt.
- Vulnerability is the birthplace of love, belonging, joy, courage, empathy, and creativity. When we're vulnerable, we allow ourselves to connect with others on a deeper level, experience joy and happiness, and be creative.
- Vulnerability is not optional. We all experience vulnerability, whether we choose to acknowledge it or not. The question is not whether we'll be vulnerable, but how we'll choose to respond to it.
- Vulnerability is the path to a wholehearted life. When we embrace vulnerability, we open ourselves up to a richer, more meaningful life.

Brown's work on vulnerability has had a profound impact on the way we understand human connection and personal growth. She has inspired millions of people to embrace vulnerability and live more wholehearted lives.

Personal and professional networking is essential for career success, but it's important to remember that trust is a two-way street. It can take years to build trust with someone, but it can be lost in seconds. Here are some cautionary tales:

The Oversharer

You've just met someone new at a networking event, and they're telling you about their personal life in detail. You're starting to feel uncomfortable, but you don't want to be rude, so you listen politely, but you're wondering why they're telling you all of this.

Oversharing can be a red flag. It shows that the person doesn't have good boundaries, and it makes you wonder if they're trustworthy. If someone is oversharing with you right away, it's best to keep your distance.

The Backstabber

You've been friends with a colleague for years, and you've always trusted them. But one day, you hear that they're going around spreading rumors about you. You're shocked and hurt, and you don't know what to do.

Backstabbers are everywhere, and they can be very damaging to your career. If you find out that a friend or colleague has been backstabbing you, it's important to distance yourself from them. You don't want to be associated with someone who is untrustworthy.

The Credit Taker

You've been working on a project with a team, and you've been doing most of the work. But when it comes time to present the project, your team member takes all the credit.

Credit takers are another type of person to be wary of. They're the ones who always try to make themselves look good, even at the expense of others. If you're working with a credit taker, be sure to document your contributions and keep your own records.

The Gossip Spreader

Your coworker tells you a juicy piece of gossip about your boss. You're tempted to listen, but you know that gossip can be very damaging.

Gossip spreaders are toxic. They're the ones always spreading rumors and talking behind people's backs. If you find out that a coworker is gossiping about you, it's important to confront them about it. You also need to let your boss know what's going on.

How to Protect Yourself

The best way to protect yourself from untrustworthy people is to be careful about whom you share your information with and whom you trust. Here are a few tips:

- Do not overshare: It's okay to share some personal information with people you trust; but be careful about sharing too much. Avoid talking about your personal life or your finances with people you don't know very well.
- Trust your gut: If you have a bad feeling about someone, trust your gut. It's usually right.
- Be careful about who you trust: Don't just trust someone because they're your friend or colleague. Take your time to get to know someone before you trust them with important information.

Building trust takes time, but it can be lost in seconds. It's important to be careful about who you share your information with and who you trust. By following the tips above, you can protect yourself from untrustworthy people and build a strong network of trusted colleagues and friends.

How to Build Trust

Building trust takes time and effort. There is no quick fix. However, there are a number of things we can do to build trust with others:

- Be honest and authentic. This means being true to yourself and your values, even when it is difficult. It also means being honest with others, even when it is uncomfortable.
- Be reliable and dependable. This means keeping your promises and commitments. It also means showing up when you say you will and being there for others when they need you.
- Be respectful and considerate. This means treating others with kindness and respect, even when you disagree with them. It also means valuing their opinions and feelings.
- Be supportive and helpful. This means being there for others when they need you and offering your help and support without being asked.
- Be forgiving. Everyone makes mistakes. If someone breaks your trust, be willing to forgive them if they are truly sorry.

"Trust Is Like A Mirror Once Broken, You Never Look At It The Same Way Again"

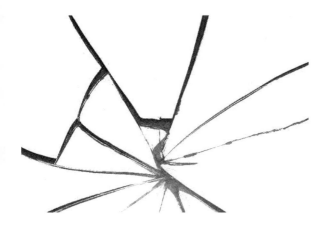

How to Repair Trust

If trust has been broken, it is important to take steps to repair it. This can be a difficult process, but it is possible. Here are a few tips:

- Acknowledge the mistake. The first step to repairing trust is to acknowledge that you made a mistake. Take responsibility for your actions and apologize sincerely.
- Explain your actions. Once you have apologized, take the time to explain your actions. This will help the other person to understand why you did what you did and see that you are truly sorry.
- Make a plan to change. Once you have explained your actions, make a plan to change your behavior. Show the other person that you are committed to rebuilding their trust.

- Be patient. It takes time to rebuild trust. Be patient and understanding with the other person. Give them time to heal and learn to trust you again.

Trust is essential for strong relationships. By following the tips above, you can build and repair trust with the people in your life.

Trustworthy People

Tim Cook of Apple: An Authentic Leader

Tim Cook, the CEO of Apple, stands as a beacon of authenticity in the corporate world. Openly gay since 2014, he brings his whole, unfiltered self to the helm of a tech titan, shattering the glass ceiling of both the industry and societal expectations. This isn't just about personal visibility; it's about the profound impact of genuine leadership.

Cook's openness normalizes diversity, paving the way for countless LGBTQ+ individuals to see themselves reflected in the highest echelons of power. He demonstrates that success isn't contingent on hiding one's identity, but rather on embracing it as a source of strength and perspective. This sends a powerful message of inclusion and belonging, not just within Apple, but across the globe.

Furthermore, Cook's leadership transcends performative allyship. He champions LGBTQ+ rights through concrete actions, advocating for equality at every turn. From Apple's unwavering support for marriage equality to its comprehensive benefits for same-sex couples, Cook's commitment goes beyond mere words. He embodies the belief that true progress lies in actively dismantling barriers and fostering a culture of acceptance.

In a world that often seeks to compartmentalize individuals, Tim Cook stands as a testament to the power of wholeness. He reminds us that true leadership lies not in conformity, but in bringing your authentic self to the table, imperfections and all. And in doing so, he inspires not just the LGBTQ+ community, but all of us, to embrace our unique identities and contribute our full selves to the world around us.

Lady Gaga Confronts Her Doubts and Fears

Lady Gaga, the pop icon known for her flamboyant persona and fierce confidence, has surprisingly admitted to battling imposter syndrome. This vulnerability reveals a fascinating paradox: the woman who dazzles audiences on stage grappling with self-doubt in the privacy of her own thoughts.

But Gaga's honesty isn't just relatable; it's deeply empowering. By sharing her struggle, she normalizes the experience for countless individuals who navigate similar anxieties. Imposter syndrome, often characterized by persistent feelings of inadequacy despite achievements, can be crippling, silencing even the most talented voices. Gaga's openness disarms that silence, offering a powerful message: even those who seem to have it all can grapple with self-doubt. Her authenticity has brought her legions of fans, ironically known as Little Monsters!

However, Gaga doesn't simply dwell on the darkness. Instead, she uses her platform to illuminate paths through the struggle. She advocates for mental health awareness, encouraging open dialogue about the challenges we face internally. She demonstrates that vulnerability isn't a sign of weakness, but rather a strength that allows us to connect, share our stories, and ultimately, overcome.

Gaga's journey with imposter syndrome reminds us that achievement and self-doubt can coexist. It's a testament to the human experience, a tapestry woven with triumphs and insecurities, brilliance and vulnerability. By sharing her own thread, Gaga invites others to do the same, stitching together a community of acceptance and understanding, where even the most dazzling stars can acknowledge the shadows within.

Tom Hanks the Nice Guy

Tom Hanks holds a unique position in American pop culture. Unlike many celebrities known for flamboyant lifestyles or controversial opinions, Hanks has cultivated an image of wholesomeness and trustworthiness. This reputation isn't just based on charm; it's a carefully constructed persona, reinforced by decades of film choices and public appearances.

Hanks often portrays ordinary characters thrust into extraordinary situations. From the astronaut lost in space in *Cast Away* to the widowed lawyer in *Philadelphia*, he embodies relatable struggles and triumphs. Audiences see themselves reflected in his characters, fostering a sense of connection and trust.

His characters often serve as moral compasses within their narratives. He frequently portrays characters who stand up for what's right, even in the face of adversity. This consistent theme reinforces a sense of integrity and trustworthiness in the public eye.

In a world of partisan politics and social media vitriol, Hanks has, on occasion, used his platform to promote unity and understanding. He avoids overt political stances but advocates for common values and respectful dialogue. He may not be a perfect person, but through his film choices and public persona, he has established himself as a Hollywood icon synonymous with honesty, decency, and the enduring appeal of the "good guy."

Whoopi Goldberg: Bringing Her Authentic Self to Work

Choosing a single figure who embodies the triumph of bringing their full, authentic Black self to work and achieving equal treatment is a challenging task since countless individuals have paved the way across various fields. However, Whoopi Goldberg's journey stands out for its unique blend of talent, tenacity, and unwavering commitment to self-expression.

Goldberg's career trajectory is a testament to her refusal to conform to predetermined molds. From her early days as a stand-up comedian defying expectations with her raw, honest humor to her iconic role as Oda Mae Brown in *Ghost*, she has consistently challenged stereotypes and carved her own path.

Goldberg's authenticity extends beyond her artistic choices. She has openly discussed her struggles with dyslexia and alopecia, defying societal pressures to present a polished, "perfect" image. This vulnerability fosters connection and inspires others to embrace their own flaws and vulnerabilities.

Furthermore, Goldberg's advocacy for social justice and equality is woven into the fabric of her work and personal life. She has been a vocal critic of racism, sexism, and discrimination, using her platform to amplify marginalized voices and advocate for positive change.

Goldberg's impact transcends entertainment. She is a cultural icon who has redefined what it means to be a Black woman in the public eye. Her success has paved the way for future generations to embrace their full selves, imperfections and all, and strive for equality in every sphere of life.

It's important to acknowledge that Goldberg's journey hasn't been without its challenges. She has faced criticism and adversity, both for her artistic choices and outspoken nature. However, her unwavering commitment to authenticity and her dedication to using her platform for good have ultimately cemented her place as a role model and a force for positive change.

While highlighting one individual's story cannot encompass the full spectrum of Black experiences in the workplace, Whoopi Goldberg's journey serves as a powerful reminder that authenticity, talent, and unwavering commitment to justice can pave the way for a more equitable future. Her story is an inspiration not just for aspiring entertainers, but for anyone who dares to be their true selves and fight for a world in which everyone is treated with respect and dignity.

Practicing What You Preach

Here are some key strategies for cultivating authenticity in your relationships:

- Stay true to your values and beliefs: Live in accordance with your values and principles, even when it's difficult. This shows others that you have strong convictions and are not easily swayed. Yet we must always be respectful of others, even if we do not share or respect their beliefs. We need to separate the person from their beliefs sometimes.
- Be transparent about your strengths and weaknesses: Don't be afraid to admit your shortcomings. Being open about your flaws allows others to connect with you on a more genuine level.

- Express your emotions authentically: Don't suppress them. Allow yourself to feel and express your emotions in a healthy way.
- Avoid pretense and putting on airs: Don't try to be someone you're not. People are drawn to those who are genuine and authentic.
- Be comfortable with your imperfections: Accepting yourself, flaws and all, is a key component of authenticity. Don't strive for perfection; embrace your unique imperfections.
- Unique idea—authenticity journal: Track and reflect on your progress by daily journaling. This journaling practice can be a powerful tool for exploring and embracing your authentic self. Dedicate time each day to write about your thoughts, feelings, experiences, and desires. This introspective process can provide greater clarity about who you are and what you value, allowing you to present your authentic self to the world.

A Lifelong Exercise

By embracing authenticity, we open ourselves up to deeper connections and create space for vulnerability and trust. This genuineness allows us to build strong, lasting relationships enriched by mutual understanding and respect. Remember, the journey toward authenticity is a continuous one. Be patient with yourself, celebrate your unique qualities, and allow your true self to shine through in all your interactions.

Maintaining Trust: A Continuous Journey

Building trust is a crucial first step, but the true test lies in maintaining it. Trust, like a delicate flower, requires constant care and attention to flourish. Here are some key principles for safeguarding the trust you've built:

- Open and honest communication: Clear and honest communication is essential for maintaining trust. Share your thoughts and feelings openly, and actively listen to understand the other person's perspective.
- Respectful and supportive interactions: Treat others with respect and consideration, even during disagreements. Offer support and encouragement, avoiding negativity or criticism.
- Forgiveness and understanding: Everyone makes mistakes. Be willing to forgive and learn from past experiences. Remember, holding onto resentment can erode trust over time.
- Continuous effort and commitment: Building and maintaining trust is an ongoing process. It requires dedication and effort from both individuals.

Discussion Questions

For Chapter 6, consider these questions:

- How do you approach networking today, and how might you experiment with the concepts discussed in this chapter?
- Draft a handwritten note of thanks to someone. Reflect on how it feels different than a text or email.

Chapter 7: Barometers of Trust

In an age when digital connections permeate our lives and the search for purpose and spiritual meaning remains constant, navigating spaces like dating apps, horoscopes, and faith communities can feel like a trust-tightrope walk. Yet, within these often-complex spheres lies the potential for meaningful connections, personal growth, and even love. But how do we step into these realms with trust at the forefront, ensuring our interactions and exploration are grounded in authenticity and respect? Are there barometers of trust to consult?

As airplane pilots, we are accustomed to relying on instruments to guide us through unpredictable skies. But what about navigating the murkier, less charted territories of life? Can we find barometers here too?

Dating Apps: Turbulent Skies with Glimmers of Hope

Imagine swiping through profiles as you would scan the weather radar, hoping to avoid emotional storms and find clear skies. Algorithms try to predict compatibility, but like wind shear, unexpected turbulence can arise. Trusting intuition becomes crucial, like using gut feelings to navigate sudden downdrafts. However, just as a pilot wouldn't abandon a flight due to temporary turbulence, don't give up on the app after a few rough encounters. Remember, even the bumpiest rides can lead to breathtaking destinations.

Honesty's Guiding Light in the Digital Dating World

When exploring the realm of dating apps, honesty becomes the guiding light. Be sure to craft profiles that truly reflect your personality, interests, and intentions. Don't shy away from disclosing deal breakers or openly communicating your preferred style of interaction. After all, genuine connection blossoms from clarity and authenticity. Remember, actions speak louder than words.

Be mindful of your communication style, follow through on plans, and respect boundaries. Ghosting, while seemingly easy, erodes trust and creates emotional harm. Trust your gut instinct too. If something feels off, don't ignore it. Prioritize your safety and well-being, and don't hesitate to disengage from anyone who makes you uncomfortable.

Pitfalls to Avoid

Misrepresentation and deception: Online dating provides a platform for individuals to present themselves in the most flattering light, and sometimes these individuals resort to embellishments or outright deception. This can lead to disappointments and hurt feelings when expectations clash with reality. For example:

- Overwhelm and choice fatigue: The sheer volume of profiles can be overwhelming, leading to decision fatigue and a tendency to swipe left without giving potential matches a chance. This can result in missed connections.

- Superficiality and ghosting: The emphasis on physical appearance and quick judgments can lead to superficial interactions and a culture of "ghosting," where individuals disappear without explanation, leaving others emotionally invested and confused.

- Catfishing and scams: Online dating platforms are not immune to scams and catfishing, where individuals create fake personas to manipulate or defraud others. It's crucial to exercise caution and be wary of profiles that seem too good to be true.

On the other hand, online dating creates more opportunities to embrace than traditional methods of meeting and dating, including:

- Expanded horizons: Online dating breaks down geographical barriers, allowing individuals to connect with people they might never have met otherwise. This can broaden perspectives and introduce new cultural experiences.

- Compatibility matching: Many dating sites utilize algorithms to match individuals based on shared interests, values, and personality traits. This can increase the likelihood of finding compatible matches.

- Efficient communication and screening: Online communication allows for efficient screening of potential partners before investing time and energy in in-person dates. This can save time and heartache.

- Confidence building and social skills development: Engaging in online dating can help individuals overcome social anxiety and improve their communication and flirting skills. It can also boost self-confidence and self-esteem.

While neither Toby nor Samir has dated online, they have surveyed countless friends and colleagues who have shared some basic strategies for success.

Transparency and Authenticity

In a world of curated profiles and carefully crafted online personas, transparency and authenticity are paramount. Be honest about your intentions, expectations, and experiences. Present yourself authentically, showcasing your interests, values, and vulnerabilities. This openness allows for genuine connection and fosters trust with potential partners.

Effective Communication and Active Listening

Clear and consistent communication is vital online, where nuances can be lost in translation. Express yourself thoughtfully, avoiding ambiguity and misleading information. Actively listen to your matches, seeking to understand their perspectives and emotions. This creates a safe space for honest communication and builds trust over time.

Building Trust Through Shared Experiences

Virtual interactions can act as icebreakers, but true connection requires shared experiences. Engage in meaningful conversations, share your passions, and explore common interests. Consider video or voice calls to deepen the connection and establish a sense of familiarity.

Protecting Yourself and Recognizing Red Flags

Online dating requires vigilance. Be mindful of the information you share and avoid revealing sensitive details too early. Be wary of inconsistencies, excessive compliments, and unrealistic promises. Trust your gut instinct, and don't hesitate to walk away from situations that feel uncomfortable or unsafe.

Taking the Leap from Online to Offline

When trust has been established, consider transitioning to offline interactions. Meet in public spaces, engage in shared activities, and continue building your connection in the real world. Remember, trust is built over time, so be patient and respectful, and enjoy the journey of getting to know someone authentically.

Additional advice on building trust in online dating includes:

- Be mindful of your profile: While your profile is your first impression, avoid over-filtered photos and exaggerated descriptions. Showcase your authentic self through candid photos and genuine descriptions.
- Engage in meaningful conversations: Go beyond superficial exchanges, and delve into deeper topics such as your passions, values, and goals. Ask thoughtful questions and actively listen to your matches.
- Be consistent and reliable: Respond to messages promptly and follow through on plans. Consistent communication and reliable behavior build trust and demonstrate your genuine interest.
- Respect boundaries and privacy: Don't pressure your matches for personal information or push them to meet before they are comfortable. Respecting boundaries and acknowledging their privacy demonstrates your maturity and trustworthiness.
- Be patient and realistic: Building trust online takes time and effort. Don't rush into relationships and set realistic expectations for online interactions. Enjoy the getting-to-know-you process and focus on genuine connections.
- Online dating can be an exciting and rewarding experience, but building trust requires a conscious effort. By being transparent, communicating effectively, and taking steps to protect yourself, you can navigate the digital world with confidence and

cultivate meaningful connections that may blossom into something special.

In a world often characterized by noise and chaos, genuine human connection remains a powerful force. By nurturing the pillars of empathy, logic, and authenticity, we can build and maintain trust in our relationships, laying the foundation for deeper connections, greater resilience, and a more fulfilling life.

As Maya Angelou eloquently stated, "People will forget what you said, people will forget what you did, but people will never forget how you made them feel." Let us strive to build trust with the intention of creating lasting positive imprints on the hearts and minds of others.

Horoscopes: Celestial Compasses, Not Autopilots

Think of horoscopes as astrological weather reports, offering insights into potential emotional currents. Like weather forecasts, they're not foolproof, but they can nudge you toward awareness. A pilot wouldn't blindly follow a faulty weather report, and neither should you rely solely on horoscopes. Use them as prompts for self-reflection, not scripts for your life. Remember, you're the pilot, not the stars.

Horoscopes can be captivating, offering glimpses into our personalities and potential futures. However, it's crucial to approach them with a healthy dose of skepticism. Remember, they are entertainment, not gospel. While they can spark inspiration and offer insights, basing life-altering decisions solely on astrological predictions can be detrimental.

Instead, use horoscopes as a tool for self-reflection, prompting you to explore your strengths, weaknesses, and hidden potential. Analyze them with a critical eye, questioning any biases or stereotypes they might perpetuate. Remember, you are the architect of your own destiny, and horoscopes, while potentially guiding, should never dictate your decisions.

In a world in which connections shape destinies, individuals often seek guidance from the celestial realm to navigate the intricate tapestry of their lives. Horoscopes, with their mystical allure, have become a unique lens through which people attempt to decipher the cosmic clues that may lead them to success, love, and personal fulfillment. However, beyond the realm of the zodiac, there lies a fascinating intersection between horoscopes and networking—a cosmic dance where the dots of destiny connect.

Astrology, often dismissed as mere pseudoscience, has endured through centuries, capturing the imagination with its promise of insight into the future. The zodiac, divided into twelve distinct signs, mirrors the diversity of personalities in the world. Each sign is associated with specific traits, strengths, and challenges, creating a celestial blueprint that guides individuals on their life journey.

Similarly, in the realm of networking, individuals navigate a vast and diverse landscape of connections. From colleagues to mentors, friends to business partners, the network we cultivate mirrors the multifaceted nature of the zodiac. Like the planets influencing our astrological charts, the people with whom we connect can shape our paths and influence our destinies.

The art of networking, akin to decoding the language of the stars, involves recognizing patterns and making meaningful connections. Just as an astrologer studies the alignment of celestial bodies to reveal hidden meanings, a skilled networker identifies opportunities and synergies that may not be immediately apparent. It's about connecting the dots between people, industries, and ideas to create a constellation of possibilities.

The zodiac teaches us that each sign has its strengths and weaknesses, just as each person in our network brings unique skills and perspectives. A successful networking strategy involves understanding these nuances, leveraging the strengths of each connection, and navigating challenges with grace—much like balancing the cosmic energies within an astrological chart.

Moreover, just as astrologers believe that the alignment of stars influences our destiny, networking often relies on being in the right place at the right time. Attending events, joining professional groups, and engaging in online communities serve as the constellations of the networking universe. It's in these spaces that the dots align, providing opportunities for collaboration, mentorship, and career advancement.

In essence, horoscopes and networking share a common thread—the belief that our destinies are shaped by the connections we make. Whether consulting the stars for guidance or attending a networking event for career prospects, both endeavors involve a certain level of faith in the unseen forces at play.

As you navigate the cosmic currents of your life, consider the parallels between horoscopes and networking. Connect the dots between your celestial chart and your professional circles, recognizing that the universe may have a plan for you—one that unfolds through the meaningful connections you create. After all, in the vast tapestry of existence, the celestial and professional realms are intricately woven together, waiting for you to connect the cosmic dots.

Faith Communities: Finding Calm Amidst the Storm

Faith communities can be sanctuaries offering solace and support, like finding clear skies within a storm system. They provide anchors of shared values and rituals, like navigational beacons guiding you through life's tempests. But just as you encounter different weather patterns, be open to exploring different faith communities to find one that resonates with your own internal compass. Remember, the goal is to find a community that uplifts you and does not restrict your flight path.

Trust in God: The Hidden Pilot in Your Network

Finding a faith community aligned with your values and beliefs can be akin to discovering a second home. Seek communities where you feel comfortable expressing yourself openly and honestly, engaging in respectful dialogue, even when faced with differing opinions. Active listening and striving to understand diverse perspectives are crucial in fostering trust and belonging. While shared values provide a foundation, respect for boundaries is paramount. Don't pressure others to conform to your beliefs or expectations. Embrace the richness of diversity and celebrate the shared values that bind your community together.

In the bustling marketplace of life, where we navigate the intricacies of personal and professional networks, the question of trust looms large. We curate connections, exchange cards, and seek alliances, but often, a deeper sense of security remains elusive. This is where for many of us, trust in God, a concept often relegated to the realm of the spiritual, can offer a surprising compass for navigating the social landscape.

Think of your network as a complex web of relationships, each thread anchored in trust. We rely on colleagues for expertise, friends for support, and mentors for guidance. But what happens when the ground beneath our connections feels shaky? When uncertainties cloud our judgment and disappointments erode trust? It's in these moments that faith in a higher power can serve as a stabilizing force, guiding us through the labyrinth of human interactions.

Here's how trusting in God can enhance your networking experience:

- Courage to connect: Fear of rejection can paralyze us, hindering the very act of building connections. Trust in God's bigger plan, however, can embolden you to step outside your comfort zone. Remember, He has placed you in this unique network for a reason, and He will equip you with the tools and courage to navigate it.
- Openness to unexpected connections: Networking often feels like a calculated game, but when you trust in God's guidance, you open yourself to serendipitous encounters. Be receptive to unexpected introductions, seemingly random conversations, and unfamiliar paths. You might discover connections that align with your divine purpose, leading to unexpected collaborations and fulfilling relationships.

- Resilience in the face of challenges: Even the most robust networks are not immune to challenges. Misunderstandings, conflicts, and betrayals can leave us feeling disillusioned and isolated. In these moments, remember that God's presence is constant. Lean on your faith to find strength, forgiveness, and the ability to rebuild trust, both with yourself and with others.
- Humility in the pursuit of success: Ambition is crucial for professional growth, but unchecked, it can breed arrogance and disconnect us from our true purpose. Trust in God reminds us that success is not a solo journey but a collaborative effort. It fosters humility, allowing us to see ourselves as part of a larger divine plan and value the contributions of others in our networks.

Ultimately, trust in God is not a replacement for initiative or effort. It can be a silent pilot guiding your navigation, offering a sense of peace and purpose amidst the uncertainties of networking. It can allow you to connect with authenticity, build relationships with trust, and navigate challenges with resilience, all while recognizing your place in the larger tapestry of His plan.

So, as you navigate the ever-changing landscape of your network, remember the invisible hand of faith may be guiding your way. Trust in His wisdom, embrace the unexpected, and let your network become a testament to the power of divine connection in the human experience.

Life's Barometer: Trust Yourself

Ultimately, the most reliable barometer for navigating these uncharted territories is your own intuition. Just as a pilot constantly monitors instruments and adjusts course, be mindful of your emotions and values. Don't be afraid to make course corrections if something feels off. Trust your gut, learn from your experiences, and remember, even the most challenging journeys can lead to beautiful discoveries.

So, the next time you find yourself swiping through profiles, deciphering horoscopes, or looking to God for answers, remember the pilot's perspective. Use the available tools for guidance; but trust your own internal compass to navigate the skies of love, faith, and life itself. After all, the smoothest flight is often the one where you're in control.

Trust: A Journey, Not a Destination

Building trust is a two-way street. While extending trust to others is important, safeguarding your own boundaries is equally crucial. Be open to vulnerability; but share information only when comfortable. Remember, trust is earned, not given freely. Take time to get to know people and communities before fully entrusting them. Observe their actions, their values, and how they treat others. Most importantly, remember that trust is a journey, not a destination. It takes time, effort, and a willingness to forgive mistakes and learn from experiences.

By approaching these spaces with intention, honesty, and respect, we cultivate trust and create deeper, more meaningful connections in our pursuit of love, guidance, and belonging. Remember, trust is the foundation of healthy relationships, both online and offline, and it's worth nurturing in every interaction. As we navigate the intricate landscapes of dating apps, horoscopes, and faith communities, may trust be our guiding compass, leading us toward genuine connections and enriching experiences.

Remember, networking is not just about collecting cards, it's about building bridges. And with trust in yourself, your family and friends, your horoscope, and even God as your compass, those bridges can lead you to unexpected destinations, meaningful connections, and a sense of belonging that transcends the limitations of the earthly marketplace.

Discussion Questions

For Chapter 7, consider these questions:

- Do you rely on any of the barometers discussed in this chapter, and if so, how are they helpful? What are their limitations?
- When have you had the greatest success in connecting with strangers, and why do you think it happened? Are their lessons to be gleaned from your past success?
- What advice do you give others about dating or making friends in our digital era?

Chapter 8: Cross-Check: Nurturing Your Relationships As You Navigate Life

As pilots, we literally take our lives into our own hands. And if we have passengers on board, our responsibility grows.

We are reminded of this responsibility every time we file a flight plan and note the number of "souls on board." This is a term used to describe the total number of passengers and crew on an aircraft. This includes all living bodies on board, such as pilots, flight attendants, crew members, and passengers. The term "souls" is even used by pilots when declaring an emergency so that rescuers know how many people to search for.

This stark fact underscores our responsibility as a pilot in command (PIC). Missteps can be deadly.

While it may sound hyperbolic, we advise exercising a similar sense of responsibility in growing your personal and professional relationships. This diligence is key to the legacy you will leave.

Just as a pilot relies on a network of instruments to keep the plane aloft, our personal and professional relationships thrive on a constant feedback loop. We monitor the emotional gauges—the warmth in a loved one's smile, the tension in a colleague's voice, the unspoken hurt in a friend's silence. These subtle cues, like altimeter readings, tell us where the relationship stands and whether adjustments are needed.

Building and maintaining meaningful networks requires intentional effort and a genuine desire to connect with others. Notable experts and leaders emphasize the importance of setting clear goals and targeting the right audience. For example, defining whether your aim is to expand your client base, find a mentor, meet new neighbors, share a hobby with another enthusiast, or explore new career opportunities, can streamline your networking efforts and lead to better results. Gary Vaynerchuk, an entrepreneur and author, highlights that networking is about building relationships rather than just making connections.

Authenticity and sincere interest are crucial in networking. Brené Brown, who we have mentioned earlier, is a research professor and author who defines authenticity as embracing who we are instead of who we think we're supposed to be. Michelle Obama, former First Lady, suggests being genuinely interested in others' stories by asking thoughtful questions and listening actively. This approach builds trust and fosters deeper connections. The answers to the questions will also surprise you and engage you more than you might have expected. Keith Ferrazzi, author of "Never Eat Alone," emphasizes that the real currency of networking is generosity, not greed, encouraging individuals to offer value without expecting anything in return.

Following up and nurturing relationships is essential for maintaining strong networks. Harvey Mackay, a businessman and author, compares a networking relationship to a plant that needs nurturing to thrive. Personalized messages, expressions of appreciation, and periodic check-ins can keep connections alive.

Being unforgettably helpful and focusing on what you can do for others are fundamental principles of effective networking. Adam Grant, an organizational psychologist and author, advises offering help without expecting anything in return, as it often comes back tenfold. Sharing knowledge, resources, and connections generously can make you unforgettable to others. Tim Sanders, author of "Love Is the Killer App," believes in treating every contact as a potential lifelong relationship, emphasizing the value of long-term connections.

Expanding others' networks by making introductions and connections is a powerful way to help. Michele Jennae, a networking expert, notes that networking is about connecting people with people, ideas, and opportunities. Introducing individuals in your network who could benefit from knowing each other can be one of the most valuable actions you take. Adam Rifkin, an entrepreneur and author, encourages being a connector by facilitating beneficial introductions within your network.

Ariana Huffington, co-founder of The Huffington Post, suggests attending industry events, conferences, and meetups to connect with like-minded individuals. By combining these strategies, you can build a vibrant, supportive network that offers mutual benefits.

Consider these examples as guidance on how one goes about collecting interesting people in their lives and maintaining those relationships over time. Social media makes people think they have a larger network than they do because it is easy to connect and like; but it does not equate to a traditional relationship developed over time. Think of relationships as ever evolving and experiment with a variety of means to stay connected with contacts across time. If you find it harder to do, then consider letting go, but only after you have reflected fully on the reasons why the relationship no longer merited your attention. As Toby likes to say, be sure that you are advancing the narrative with your network. Reunions and past experiences are fine to acknowledge when you regroup with people from your past; but be sure to create present and future topics and activities to connect over so that your relationships continue to grow with you. It has the added benefit of also keeping you younger and more engaging than those who do not.

But focusing solely on these internal gauges, like a pilot fixated solely on the instrument panel, can lead to disaster. Just as a pilot must scan the horizon for storms and turbulence, we must also be mindful of external factors impacting our relationships. The "air currents" of life—work stress, family tensions, financial pressures—can all affect how we interact and communicate. Ignoring them can lead to misunderstandings, resentment, and ultimately, a crash landing.

Let's Explore Instruments of Connection:

- Altimeter: Checking your "altitude" in a relationship involves assessing its overall health. Are you communicating openly? Do you feel supported and appreciated? Are you both growing and evolving together?
- Airspeed: The "airspeed" of a relationship refers to the pace and frequency of interaction. Are you making regular efforts to connect, share experiences, and offer support? Are you finding a comfortable balance between closeness and independence?
- Heading indicator: This represents the shared direction and goals of the relationship. Do you have common values, aspirations, and dreams for the future? Are you working together to achieve them?
- Fuel gauge: The emotional "fuel" of a relationship is built on trust, respect, and affection. Are you making consistent deposits through acts of kindness, empathy, and understanding? Are there any "leaks" of negativity, criticism, or resentment draining the tank?
- Horizon indicator: This signifies the ability to navigate challenges and conflicts. Can you communicate effectively, compromise when needed, and emerge stronger after disagreements?

Cross-Checking in Action:

- Regular check-ins: Just as a pilot scans their instruments every few seconds, make regular "check-ins" with your loved ones. Schedule time for meaningful conversations, ask questions, and actively listen to their concerns and joys.
- Course corrections: If an instrument indicates a problem, the pilot takes corrective action. Similarly, if you sense something is amiss in your relationship, address it openly and honestly. Communicate your needs, apologize for mistakes, and work together to find solutions.
- Trusting in autopilot: While constant vigilance is crucial, a good pilot also knows when to trust the autopilot. Similarly, in healthy relationships, trust and understanding can provide a foundation of stability, allowing for occasional "autopilot" moments where things flow naturally without needing constant monitoring.
- Weathering storms: Just as a pilot navigates turbulence and unexpected weather, relationships also face challenges and storms. Stay calm, communicate openly, and work together to weather the difficulties. Remember, even the strongest storms can make the flight path clearer and the bond stronger.
- Celebrating the journey: Just as a successful landing is celebrated, cherish the milestones and achievements in your relationships. Express gratitude for your loved

ones, celebrate their successes, and enjoy the journey together.

Beyond the Analogy

While the pilot analogy provides a helpful framework, remember that relationships are more complex than a flight path. They are dynamic, evolving ecosystems with nuances that instruments alone cannot capture. Ultimately, maintaining strong relationships requires a blend of attentiveness, communication, trust, and a genuine commitment to nurturing the connection over time.

So embrace the role of the relationship copilot, constantly scanning the instruments of connection, making adjustments as needed and celebrating the joy of this beautiful, shared journey.

The importance of maintaining relationships cannot be overstated. The benefits are infinite, including:

- Improving our mental and physical health. Studies have shown that people with strong social connections have lower rates of depression, anxiety, and loneliness. They also tend to live longer and healthier lives.
- Providing us with support and love. Our relationships can provide us with emotional, financial, and practical support. Our loved ones can also help us through difficult times and celebrate our successes.

- Giving us a sense of belonging. We all need to feel like we belong to something. Our relationships can provide us with a sense of community and belonging.
- Helping us to learn and grow. Our relationships can help us learn new things and grow as individuals. We can learn from our loved ones' experiences, perspectives, and values.

You do not have to be a pilot to understand the importance of cross-checks and maintaining healthy relationships. In lay terms, here are a few tips:

- Make time for the people who are important to you. It is important to schedule time for your relationships. This doesn't mean that you have to spend every weekend with your friends and family; however, it does mean making an effort to connect with them regularly.
- Be present and engaged when you are with the people you care about. When spending time with someone, put away your phone and other distractions. Give the person your full attention, and listen to them attentively.
- Be supportive and helpful. Be there for your loved ones when they need you. Offer your help and support without being asked.
- Be forgiving. Everyone makes mistakes. If someone you care about hurts you, be willing to forgive them if they are truly sorry.

- Nurture your relationships. Relationships take time and effort to maintain. Make an effort to do things that you enjoy with your loved ones and create new memories together.

Life can get busy, and it can be difficult to maintain relationships over time. Don't beat yourself up if you can't always be there for a friend; but be accountable to them and your relationship. For example:

- Make an effort to stay in touch. Even if you can't see someone in person regularly, you can stay in touch through phone calls, text messages, email, or social media.
- Be thoughtful and caring. Send your loved ones cards or gifts to let them know that you are thinking of them. Digital tools like American Greetings and Google calendars make it easier to remember important dates.
- Celebrate their successes. Be supportive of your loved ones' goals and dreams. Celebrate their successes with them and be there for them during their setbacks.
- Make time for visits. If possible, try to visit your loved ones in person on a regular basis. This is a great way to reconnect and create new memories together.

Maintaining relationships is important for our well-being. By following the previous tips, you can keep your relationships strong and healthy.

False Sense of Security

Many people believe that once they've established a relationship, they can relax and let their guard down. This is a false sense of security. Relationships require constant maintenance and effort. If you don't put in the work, the relationship will eventually fade away.

Here are a few examples of how people can fall into the trap of false security in relationships:

- Thinking that a relationship is strong because it's been around for a long time. Just because two people have been together for a long time doesn't mean that their relationship is strong. It's important to continue to nurture the relationship, even after many years.
- Assuming that a partner will always be there for them, no matter what. No one is obligated to stay in a relationship, even if they've been together for a long time. It's important to appreciate your partner and make them feel loved and valued.
- Taking the relationship for granted. It's easy to take a relationship for granted when things are going well, but it's important to remember that relationships are fragile and can be easily damaged.
- Be forgiving. Everyone makes mistakes. Be willing to forgive your partner when they make a mistake.
- Show your appreciation. Let your partner know how much you care about and appreciate them.

Maintaining relationships takes time and effort, but it's worth it. Strong relationships make us happier and healthier. They also provide us with support and love, which are essential for our well-being.

Remember, just like piloting, maintaining relationships is a constant learning process. There will be turbulence, unexpected bumps, and moments of white-knuckled flying. But by embracing the dual focus of internal gauges and the ever-changing external landscape, we can navigate the skies of our relationships with grace, resilience, and a clear view of the horizon.

We can also learn from others who have mastered relationships for the long haul. In any generation, there are standout "super connectors" who rise above the ordinary, with an uncanny ability to forge strong relationships across diverse landscapes and throughout their lives. Let's first consider some of their traits and then look at specific individuals.

Super connectors share some common traits, including:

- Masterful communication: Like seasoned air traffic controllers, they listen actively, picking up subtle signals and guiding conversations with ease. Empathy flows through their headsets, connecting them to stories from all walks of life. It's not just

about hearing; it's about understanding, a vital skill for building trust and lasting partnerships.
- Altruism above all else: True superconnectors know their worth isn't measured in collected business cards, but in the impact they leave on others. They're constantly scanning the radar for opportunities to be of service, offering a helping hand like a skilled wingman to those in need. Their actions whisper a universal message: "I see you, and I'm here for you."
- Embracing the multiverse of perspectives: They understand that diversity isn't just a buzzword; it's the oxygen that fuels innovation and growth. They value different viewpoints, welcoming them into the cockpit like essential instruments, enriching their own flight paths and those around them. Learning from others becomes a constant quest, a perpetual climb to new altitudes of understanding.
- Champions of inclusivity: Their cabins are never exclusive. Super connectors actively seek out passengers from all backgrounds, understanding that the richness of connections blossoms in the fertile ground of diverse experiences. They break down barriers and bridge cultural canyons, ensuring everyone feels welcome and heard in the human skyway.

Imagine a world where a financial titan casually chats with a pop icon, an actor becomes the ultimate bridge between Hollywood A-listers, and a philanthropist convenes global leaders with a wave of their hand. This is the realm of super connectors, and within it, figures like Warren Buffet, Beyoncé, Kevin Bacon, David Rockefeller, and Barbra Streisand weave webs of connection that transcend industries and redefine influence.

Warren Buffet isn't just an investing savant; he's a maestro of relationships. He cultivates genuine connections with industry giants and young minds alike, fostering trust and loyalty that blossom into a network richer than any portfolio. He's the bridge between the boardroom and the lecture hall, ensuring knowledge flows freely and mutually beneficial partnerships bloom.

Beyoncé, on the other hand, wields social media like a magic wand. Her online community isn't just a following; it's a vibrant tapestry woven with fans and industry figures alike. Collaborations blossom organically, each one a thread strengthening the fabric of her influence. And beyond the digital realm, and her famous BeyHive, her status as a cultural icon attracts influential figures seeking to be part of her tapestry.

Kevin Bacon isn't just an actor; he's the ultimate connector in Hollywood's sprawling web. His filmography is a constellation, each project a link to countless actors, directors, and productions. He readily ventures into diverse genres, building bridges between them with every new role. Moreover, his infectious enthusiasm keeps him ever-present in industry events, mentoring young actors, and ensuring the web of connections never frays. (We will discuss Bacon's unique constellation of connections further in Chapter 10.)

David Rockefeller wasn't just a wealthy man; he was a philanthropist with a global Rolodex. His wealth and influence became tools for connection, forging bonds between individuals and organizations across sectors. He brought world leaders and everyday citizens to the same table, his network a catalyst for collaboration and positive change.

Barbra Streisand isn't just a multi-talented artist; she's a magnet for connections. Her multifaceted career has placed her at the crossroads of music, film, activism, and social causes, bringing her into contact with a dizzying array of influential figures. But Streisand goes beyond mere proximity; she nurtures long-lasting friendships and mentorships, transforming her network into a supportive family that amplifies her voice and reach.

These are just a few examples of how super connectors operate. They don't hoard connections; they cultivate them, offering value, bridging divides, and creating a powerful ecosystem where influence isn't a solo act, but a symphony of connections.
They're not just networking; they're building bridges, fostering collaboration, and reminding us that in the vast human universe, connection is the compass that guides us all toward a brighter, more connected tomorrow. So, take a leaf out of their flight log, engage your genuine engines, and open your cockpit to the diverse landscape of human interaction.

You might just discover you, too, have the makings of a super connector within.

Discussion Questions

For Chapter 8, consider these questions:

- How do you maintain your closest relationships? How frequently are you in touch, and how do you keep in touch? Are there practices you always follow or actively avoid, such as only texting versus emailing or phoning?
- How much of your relationship management is digital, and how much is analog? What might happen if you tried to shift the balance?
- Have you ever lost a relationship, and how did it happen? Conversely, have you ever rejuvenated a relationship, and how did it happen? How did it make you feel?

Chapter 9: Networking in the Digital Age

Social media has become an integral part of our lives. We use it to connect with friends and family, share news and updates, and stay informed about what's going on in the world. But what we may not realize is that social media companies use our data to predict who we will connect with in the future.

Remember in *The Wizard of Oz* when Dorothy's dog, Toto, pulls back the curtain in Oz to reveal the underwhelming wizard of Oz? Well, keep that image in mind as you navigate the digital era infused by the steroid of AI.

In this chapter, we explore how social media websites like LinkedIn and Facebook use algorithms to predict our connections. We also discuss the implications of this technology and how it could be used to manipulate our behavior.

The Algorithm Knows

Social media companies collect vast amounts of data about us, including our likes, shares, and posts. They also track our location, phone contacts, and browsing habits. This data is then fed into algorithms that are designed to predict our behavior.

One way that social media companies use this data is to predict with whom we will connect. For example, if you frequently like and share posts from a particular person, LinkedIn may suggest that you connect with them. This is because the algorithm has determined that you are likely to be interested in connecting with this person.

Social media companies also use this data to target us with advertising. For example, if you've been searching online for new cars, you may start seeing ads for car dealerships on your social media feed. This is because the algorithm has determined that you are likely to be interested in buying a car.

The Implications

The ability to predict our connections has a number of implications. On the one hand, it can be used to help us connect with people who we might not have met otherwise. For example, LinkedIn can suggest that you connect with people who work in the same industry as you or who attended the same school.

On the other hand, the ability to predict our connections can also be used to manipulate our behavior. For example, social media companies could use this data to show us news articles that are designed to make us feel a certain way. They could also use this data to target us with advertising designed to make us buy certain products.

It is important to be aware of the implications of this technology. We should be careful about what information we share on social media, and we should be skeptical of the information that we see on our social media feeds.

Work Your Technology

As Toby is fond of saying, "Work your technology, don't let it work you." As AI grows into our lives, new systems are learning autonomously and making complex judgments that we need to understand.

In December 2023, Toby's friend Henry Timms, then CEO of New York City's celebrated Lincoln Center, co-authored a Harvard Business Review article on AI and machine learning with Jeremy Heimans. Their article on the era of the "auto sapient" focuses on the role of AI in leadership. They understand how AI has been subtly influencing us for years, and they show us how a new generation of vastly more capable technology is emerging. These systems, the authors write, aren't just tools. They're actors that will increasingly behave autonomously, making consequential moves.

The authors also encourage readers to prepare for the age of the DSO—digital significant other. We interpret this as a call to train your algorithms. In other words, work your technology before it works you.

We wholeheartedly endorse the Timms and Heimans' view.

Timms and Heimans write: "A new generation of AI systems are no longer merely our tools—they are becoming actors in and of themselves, participants in our lives, behaving autonomously, making consequential decisions, and shaping social and economic outcomes."

They offer readers a way to conceptualize and navigate this new world, in and out of the office. They introduce us to the Age of Auto Sapience, where digital systems are autonomous and humanlike, possessing a type of wisdom to make complex judgments in a context that can rival that of humans.

The authors attribute four key characteristics to autosapient systems: "They are *agentic* (they act), *adaptive* (they learn), *amiable* (they befriend), and *arcane* (they mystify). These characteristics help us understand the right way to approach them and how and why they are set to wield increasing power."

As you build your networks and live your lives, now more than ever it is critical to proactively engage with AI to understand how it will influence your destiny. Timms and Heiman's chart below on the Dynamics of Power is chilling and exciting at the same time. Better that we pull back the curtain on the AI wizard to better understand what is at work and how to best engage.

The Future of Networking in a World of Autosapient Technology

The rise of autosapient technology is going to fundamentally change the way we interact with the world around us, including the way we network. Here are a few key points to consider:

- We will need to learn to collaborate with autosapient systems. These systems will provide us with valuable insights and assistance, but they will not be perfect. We will need to develop the ability to work with them effectively, both as colleagues and tools.
- There will be a premium on human connection. As technology becomes more sophisticated, there will be an even greater need for human connection. People will be drawn to those who can provide genuine empathy, understanding, and support.
- Networking will become more important than ever. In a world of constant change, it will be more important than ever to build strong relationships. These relationships will provide us with the support and resources we need to navigate the challenges and opportunities of the future. For example, if you have

doubts about your cultural understanding of a topic, you can use AI as a sounding board. For example, do you appreciate the differences in the meaning of *Asian* versus *Oriental* or *black* versus *Black*? What about *queer* versus *gay*? Considering such points is a positive thing. And if you have nobody to turn to, AI will help you understand and make the right word choice for the specific moment.

Data is Destiny

AI expert and author Joy Buolamwini and other experts in the field of artificial intelligence have stated that with AI, "data is destiny." We embrace this view and acknowledge that it is a complex and multifaceted statement, open to various interpretations and perspectives. Here are some key points to consider:

Potential Interpretations:

- Data as fuel: AI models rely heavily on data for training and generating outputs. In this sense, high-quality, diverse data can lead to more accurate and fair AI systems while biased or limited data can perpetuate existing inequalities and harm marginalized groups. Therefore, data quality can significantly influence the outcomes and impact of AI systems, shaping our future.
- Data as a reflection of reality: AI models are trained on data that represents the real world. However, this data can often reflect existing biases and injustices present

in society. If AI systems are simply replicating these biases, it can solidify them further, limiting our ability to achieve a more equitable future.
- Data as a tool for change: While data can reflect existing problems, it can also be used to identify them, challenge them, and advocate for change. By analyzing and understanding data, AI can help us uncover biases, design fairer systems, and promote positive social outcomes.

Ethical Considerations:

- Data privacy: The use of personal data for AI development raises concerns about privacy and potential misuse. Balancing data collection and usage with individual privacy is crucial.
- Algorithmic bias: Ensuring data used for AI systems is representative and free from bias is vital to avoid perpetuating discrimination and unfairness.
- Transparency and accountability: AI systems should be transparent in their decision-making processes, and individuals should be able to understand and challenge biased outcomes.

Overall, the concept that data is destiny highlights the crucial role data plays in shaping the development and impact of AI systems. It serves as a reminder of the importance of responsible data collection, analysis, and usage to ensure AI contributes to a just and equitable future.

The "destiny" aspect should not be interpreted as a deterministic prophecy, but rather a call for responsible action. We have the power to influence the direction of AI development by making informed choices about data usage and ethical practices.

Open discussions and collaborations between various stakeholders, including technologists, policymakers, and the public are essential to ensure AI serves humanity in a positive way.

We encourage you to explore this topic further and form your own opinion on the relationship between data, AI, and our collective destiny.

Joy Buolamwini did just that when she created the Algorithmic Justice League.

In 2016, a TED Talk by Buolamwini sparked a revolution. Armed with research revealing racial bias in facial recognition technology, she founded the Algorithmic Justice League (AJL), determined to ensure that AI doesn't perpetuate harm against marginalized communities.

Starting as a one-woman crusade, the AJL quickly gathered momentum. Buolamwini's unique approach, "rounding" technical analysis with impactful storytelling through art and multimedia, proved powerful. Projects like "Gender Shades" exposed facial recognition bias through research, while "Coded Bias" brought the human impact to life through personal narratives.

This multifaceted approach has become the AJL's hallmark. It delves deep into AI's implications in areas like healthcare, criminal justice, and education, publishing rigorous research while using documentaries and art installations to raise public awareness. The AJL doesn't just point out problems; it empowers communities. Initiatives like "Community Reporting of Algorithmic System Harms" give individuals the tools to fight for AI justice themselves.

The AJL's relentless efforts have resonated globally. The AJL has influenced policy discussions, collaborated with tech giants, and earned prestigious awards. Today, it is a leading voice in the fight for fair and accountable AI, recognized for its ability to "round" out the conversation, ensuring technology serves all communities equally.

The AJL's journey is far from over. It continues to expand its research, advocate for legislation, and empower communities. So whether you're a researcher, activist, or simply concerned about the ethical implications of AI, join the movement. Visit the AJL's website, dive into its work, and become part of the solution. Together, we can ensure AI serves humanity and does not divide it.

What Now?

Kara Swisher is an American journalist and the author of numerous books. Her most recent memoir, *Burn Book: A Tech Love Story*, was published in February 2024.

In a February 26, 2024 interview with NPR, she says this about technology.

"I had great hopes, and I still do. Like, when I think of artificial general intelligence, there's a lot of scary things, but I think of all the great things. I always tend to go toward, 'What could this do? What could we do for education?' I have a real obsession with talent, where talent is. And I always think that one of the great things about tech is you can find talent anywhere. Before, it was trapped in, I don't know, a little girl in Syria that couldn't get education, well, now she can. She can be connected to the wider world. I always believed that connection brings better outcomes because if people could see their commonality. What it's done because of the way it's been rolled out has fractured us, isolated us, and made us not understand each other as well."

According to Swisher, technology fractures and isolates us because it wasn't rolled out in a way that brings people together. Swisher argues that tech leaders have not lived up to their promises to flatten organizations and bring education to everybody. Instead, the way tech has been implemented has created a situation in which people are connected but not communicating. We need to engage with AI to ensure that this does not remain the trend line.

Considerations for Social Media and Phone Usage for Kids

In an *Atlantic* article by Jonathan Haidt, "End the Phone-Based Childhood Now. The Environment in Which Kids Grow up Today Is Hostile to Human Development," Haidt makes a compelling case for delaying giving children smartphones.

Haidt's research delved into the effects of smartphone usage on the mental health of teenagers, exploring the correlation between increased screen time and a decline in psychological well-being. It highlights a recent study published in the journal *Clinical Psychological Science*, which suggests that adolescents who spend more time on smartphones are more likely to experience symptoms of depression and loneliness.

We second Haidt's argument that some technology works to the detriment of meaningful offline connections which is a contributing factor for the rise of anxiety and depression in our society. He is focused on youth, but we and many of our peers would argue that all of society is impacted by this wave of reduced offline connection. We believe that our online connections offer a false sense of relationship, which works against all of the concepts that we describe in this book.

Researchers have identified several potential factors contributing to this phenomenon, including the displacement of face-to-face social interactions, the addictive nature of social media platforms, and the constant comparison with others' curated online lives. Additionally, excessive screen time can disrupt sleep patterns, further exacerbating mental health issues.

However, the article also acknowledges the complexity of the relationship between smartphone use and mental health, noting that not all adolescents are equally affected. Factors such as socioeconomic status, family dynamics, and individual resilience play significant roles in determining susceptibility to negative outcomes.

The findings underscore the importance of promoting healthy digital habits and fostering meaningful offline connections among teenagers. Educators, parents, and policymakers are encouraged to implement strategies that balance the benefits of technology with the need for face-to-face interactions and mental well-being. By addressing these issues proactively, society can better support young people's holistic development in an increasingly digital world.

Legislation

In an era when technology shapes the very fabric of our interactions, safeguarding online users has become paramount. The landscape of digital connectivity is evolving rapidly, prompting legislative bodies worldwide to enact measures ensuring the safety and privacy of individuals navigating the digital realm. One significant legislative stride is the Digital Services Act (DSA) proposed by the European Union.

The DSA represents a pioneering effort in the establishment of comprehensive regulations for digital platforms, aiming to foster a safer online environment while preserving the fundamental rights of users. Among its provisions are stringent measures to combat illegal content, including hate speech, terrorist propaganda, and counterfeit goods. Moreover, the DSA introduces transparency obligations, compelling platforms to disclose their algorithms' functioning and mitigate the dissemination of harmful content.

Beyond the European Union, countries worldwide are recognizing the imperative of regulating technology to safeguard their citizens. From stringent data protection laws such as the General Data Protection Regulation (GDPR) to emerging initiatives addressing the ethical implications of artificial intelligence (AI), governments are proactively shaping the digital landscape.

Here's a glimpse into how legislators and businesses are taking the lead:

Government Initiatives:

- United States: While lacking a comprehensive federal law, the US government is actively involved in shaping AI governance. The National Institute of Standards and Technology (NIST) is developing a voluntary AI Risk Management Framework that outlines best practices for responsible development, deployment, and use of AI systems.
- South Korea: The South Korean government established its own AI Ethics Commission in 2018. The commission focuses on developing ethical guidelines for AI development and use, promoting transparency and fairness in algorithms, and encouraging public participation in AI governance.
- Singapore: Singapore's Model AI Governance Framework focuses on four key areas: fairness,

accountability, transparency, and explainability (FATE) principles for AI development. This framework aims to ensure trustworthy AI systems and promote public trust in technology.
- Canada: Canada has released its Algorithmic Transparency and Accountability Framework, which focuses on building public trust in AI and ensuring that algorithmic systems are deployed fairly and transparently.

Business Leadership:

- Partnership on AI: This multi-stakeholder initiative brings together leading technology companies, research institutions, and civil society organizations to develop best practices for responsible AI development. The Partnership has published several reports and recommendations on topics such as fairness, transparency, and accountability in AI.
- Tech giants taking charge: Major technology companies such as Google, Microsoft, and IBM have developed their own ethical principles for AI development. These principles cover areas like fairness, bias mitigation, and transparency in algorithms.

The regulation and oversight of AI is a complex and evolving field. While the EU's GDPR and corporate undertakings are significant steps, it's encouraging to see more governments and businesses worldwide taking a proactive approach to responsible AI development. Collaboration and open dialogue between stakeholders will be key in shaping a future where AI benefits society as a whole.

As we venture into the future, it's foreseeable that similar legislative endeavors will gain traction globally. The convergence of technology and human connectivity necessitates a proactive approach to address emerging challenges, from privacy concerns to algorithmic biases. By instituting forward-thinking laws, policymakers can empower individuals to navigate the digital world with confidence, fostering a harmonious balance between technological advancement and human welfare.

Discussion Questions

For Chapter 9, consider these questions:

- How do you consider algorithms when you are online? Do you disregard them completely, or do you try to manage them? How?
- Have you used generative AI such as ChatGPT or Gemini? What concerns, if any, have you had when using AI?
- Are you banned from using AI at your place of work, school, or other organization? Does this put you at a competitive disadvantage?
- At what age do you believe a young person should be able to interact knowingly with generative AI such as ChatGPT or Gemini? Why?

Garmin 'Glass Cockpit'

Chapter 10: Tools in Your Toolbox

In the realm of networking tools, envision your strategies as a pilot navigates the skies of professional connections. Just as an aviator relies on a diverse set of instruments to ensure a smooth flight and reach their destination, professionals utilize various apps as indispensable tools in their networking toolbox.

You might think of your smartphone as the cockpit, equipped with essential applications that act as the flight controls for your networking journey. Each app serves a specific purpose, akin to the distinct instruments a pilot relies on to maintain course, altitude, and communication during a flight.

Just as an altimeter gauges altitude, consider LinkedIn your professional altitude indicator, helping you gauge and elevate your career standing. Twitter, functioning like a communication radio, keeps you connected to real-time industry updates and conversations, allowing you to navigate the ever-changing landscape of your field.

Instagram, the panoramic window of your networking aircraft, provides a visual narrative of your professional journey, allowing others to see your trajectory and interests. TikTok, on the other hand, acts as the agile maneuvering system, enabling you to creatively navigate through challenges and trends in the professional airspace.

The direct messaging features of these apps serve as your communication radio, facilitating direct contact with other professionals. Hashtags and geotags act as navigation waypoints, guiding you to specific communities and locations where you can connect with like-minded individuals.

Just as a pilot selects the right combination of instruments for a successful flight, strategically choosing and utilizing these networking apps ensures a well-rounded approach to professional relationship building. So, fasten your seatbelt, embrace your networking cockpit, and soar through the interconnected skies of your career with the perfect set of tools in your toolbox.

Like the pilot on the ground adjusting their instruments before takeoff, the altimeter is of the utmost importance. It indicates how far above sea level you are, and once airborne, the reading will fluctuate based on location, atmospheric conditions, and the pilot's own state. In a similar sense, figuring out just how big, diverse, and current your network is will be critical to growing it. Actor Kevin Bacon is a great case study in taking stock of one's network.

Image used under license from Shutterstock.com

Six Degrees of Kevin Bacon: More Than Just a Game, It's a Glimpse into Humanity

Kevin Bacon, the man with the impossibly charismatic smile and a filmography that seems to stretch into infinity, has become synonymous with a unique phenomenon: the Six Degrees of Kevin Bacon. This playful game, which posits that any actor can be connected to Bacon within six film appearances, has transcended its entertainment roots to become a fascinating exploration of interconnectedness.

The game is directly related to the Broadway play Six Degrees of Separation by John Guare, which premiered in 1990.

In the play, one of the characters, Ouisa Kittredge, mentions the concept of "six degrees of separation" -- the idea that any two people on Earth are separated by at most six social connections. This concept inspired the creation of the parlor game.

At its core, Six Degrees of Kevin Bacon is a celebration of collaboration. It reminds us that even the most celebrated stars are not lone wolves, but rather nodes in a vast network of creativity and shared experiences. Each film appearance becomes a bridge, a connection forged through shared stories and cinematic magic.

Some examples of people Kevin Bacon is connected to within six degrees:

- Tom Hanks: Kevin Bacon starred in *Apollo 13* (1995) with Tom Hanks.
- Meryl Streep: Tom Hanks starred in *The Post* (2017) with Meryl Streep.
- Leonardo DiCaprio: Meryl Streep starred in *Doubt* (2008) with Philip Seymour Hoffman.
- Daniel Day-Lewis: Philip Seymour Hoffman starred in *Gangs of New York* (2002) with Daniel Day-Lewis.

- Nicole Kidman: Daniel Day-Lewis starred in *Nine* (2009) with Nicole Kidman.
- Hugh Jackman: Nicole Kidman starred in *Australia* (2008) with Hugh Jackman.

A website called The Oracle of Bacon can be used to calculate the number of degrees of separation between any two actors. As well, the Six Degrees of Kevin Bacon game has been the subject of several academic studies, which have found that it is actually a fairly accurate representation of the film industry's interconnectedness.

But the game's appeal extends beyond Hollywood. It taps into a fundamental truth about human nature: we are all interconnected, woven together by a tapestry of relationships, experiences, and shared history. The six degrees between us might not be measured in film appearances but in shared friends, distant relatives, or even chance encounters.

The game serves as a reminder that we are not islands and that our paths, however seemingly disparate, may have crossed in ways we never imagined. It encourages us to look beyond the surface and seek out the hidden connections that bind us together.

It also highlights the surprising power of chance—the twists and turns of fate that lead us to meet certain people and collaborate on projects, often defying logic and the odds. It reminds us that life is rarely linear and that the most unexpected encounters can hold the seeds of remarkable opportunities.

Furthermore, the game carries a valuable lesson about perspective. By stepping outside our own circles and exploring the connections between others, we gain a broader understanding of the world and the diversity of human experiences. We learn to appreciate the value of different backgrounds, skills, and perspectives, fostering empathy and understanding.

In an age of increasing polarization and division, Six Degrees of Kevin Bacon offers a refreshing antidote. It reminds us that despite our differences, we are all part of the same human story, connected by invisible threads that run deeper than we might imagine.

So, the next time you find yourself in a conversation, don't just talk about the weather. Throw out a random topic that you care about and see where it takes you. You might be surprised at the connections you discover, the stories you hear, and the sense of shared humanity that emerges.

How to Assess Your Personal and Professional Networks

Assessing your personal and professional networks can be both insightful and valuable for optimizing your connections and achieving your goals. Here's a guide to quantitatively and qualitatively assessing your network:

Quantitative assessment:

- Size and diversity: Count the total number of connections you have. Look for diversity in areas such as industry, profession, geography, age, and experience. A diverse network opens doors to various opportunities and perspectives.
- Engagement: Analyze how often you interact with your network. Track communication frequency, response rates, and the nature of your exchanges (informational, supportive, collaborative). Active engagement strengthens bonds and increases the value of your network.
- Reach and influence: Identify prominent individuals in your network. Consider their positions, achievements, and areas of expertise. Connecting with influential people can offer access to valuable resources and knowledge.
- Platform distribution: Assess where your network lives. Check online platforms like LinkedIn and social media, your contact list, and any professional associations to which you belong. Understanding your platforms will help you tailor your engagement strategies.

Qualitative assessment:

- Strength of relationships: Consider the depth and quality of your connections. Do you have mentors, trusted advisors, reliable collaborators, or close

friends in your network? Strong relationships offer personal and professional support.
- Mutual benefit: Evaluate the reciprocal nature of your connections. Do you provide and receive value equally? Networking is best when it's a two-way street.
- Alignment with goals: Analyze how your network aligns with your personal and professional aspirations. Do your connections have expertise, resources, or connections relevant to your goals? A strategically aligned network can accelerate your progress.
- Positive vibes: Evaluate the overall energy and tone of your network. Are your interactions encouraging, motivating, and inspiring? A positive network can significantly impact your well-being and drive.

Tools and techniques:

- Mapping tools: Use online tools like LinkedIn's My Network or dedicated apps like Covet to visualize your connections and identify gaps.
- Networking logs: Track your interactions with your network, noting the date, type of communication, and key takeaways.
- Surveys and feedback: Ask your connections about their perception of your relationship and how you can be more valuable to each other.

- Reflection exercises: Spend time reflecting on your network's strengths, weaknesses, and potential for growth. Identify areas for improvement and set goals for future engagement.

Remember, a great network is not just about numbers but about meaningful connections. By using both quantitative and qualitative assessments, you can gain valuable insights into your network's true potential and optimize it for personal and professional success.

AssessYourNetwork.Com

One digital tool to consider in assessing your networks is literally called AssessYourNetwork.com. It is a great tool to understand your network of friends and family and see how they interconnect.

As the website points out, your network has a huge impact on your life. Everything from your likelihood of being promoted to your happiness can be impacted by the structure of your network.

But, in our experience, most people have no idea what their network looks like. In order to strengthen your network, you must first understand it. AssessYourNetwork.com and its accompanying customer report describe three types of networks.

- Expansionists have very large networks, are popular and influential, and have an uncanny ability to work a

room. They create value by connecting contacts to each other, and they are masters at cultivating and utilizing their weak ties. Expansionists are at risk of generosity burnout as they manage large numbers of contacts.
- Brokers have diverse networks, with contacts in many different networks that do not overlap. They create value by ferrying information and identifying opportunities for collaboration between networks but need to manage misunderstandings that can arise between people with different ideas and beliefs. Brokers are innovative, often follow atypical career paths, and report better work-life balance.
- Conveners have dense, closed networks of interconnected contacts. They often live and/or work at the same location for many years, are adept at taking the perspective of others, show greater resilience, and get trust, emotional support, and buy-in from their networks. Convener networks are at risk of developing echo chambers that validate one widely held viewpoint while discounting those that oppose it. The upside of this network is that conveners are trusted and supported.

Here are some screenshots from AssessYourNetwork.com

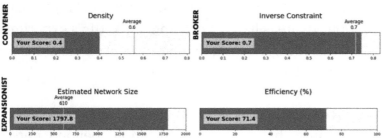

How Many People Do You Really Know?

Tyler McCormick, a professor at the University of Washington, has determined that the average person knows approximately 611 individuals. This calculation, reached through statistical analysis, offers insight not only into personal social networks but also into understanding hard-to-reach populations. McCormick's methodology involves asking individuals how many people they know with common names, such as Michael or Stephanie. By extrapolating from this data, estimates can be made about broader social circles and even marginalized groups.

While the average person may know around 611 people, there's a distinction between acquaintances and close friends. Research suggests that people typically have three or fewer close friends while maintaining a larger network of acquaintances. The sociologist Mark Granovetter's theory of The Strength of Weak Ties highlights the importance of casual connections in areas like job hunting. LinkedIn's experiments reinforce this idea, showing that peripheral connections often lead to more job opportunities than strong ties.

McCormick's method of estimating social networks based on common names has broader implications beyond personal relationships. His method has been utilized to measure hard-to-reach populations, including the homeless, by asking individuals about their acquaintances within such groups. While this approach isn't without limitations, such as verifying the accuracy of responses, it provides valuable insights into societal issues such as the prevalence of certain behaviors or identifying marginalized communities.

Despite its imperfections, McCormick's technique offers a unique perspective on society and its complexities. By understanding the breadth of social ties and the demographics within them, researchers can shed light on various social phenomena and address pressing questions. Continuous efforts to refine this method aim to make it more accurate and applicable across different demographic groups, further enhancing its utility in studying societal dynamics and challenges.

Here are the key points of McCormick's team's research:

- Problem: Traditional methods for estimating network size, like simply asking people how many people they know, are prone to underestimation and bias.
- Proposed solution: The authors developed a latent non-random mixing model that accounts for the non-uniform distribution of connections within social groups. By asking about specific sub-populations, the model can correct for biases and provide a more accurate estimate of the total network size.
- Benefits: This method is more efficient than asking about every single connection, requiring fewer questions and reducing interview time. It also produces more accurate estimates, especially for individuals with large or complex networks.
- Validation: The model is tested on real-world data and shown to be significantly more accurate than traditional methods. Additionally, it can be used to estimate the distribution of network sizes in a population.
- Implications: This research has significant implications for social network research, allowing for more accurate data collection and analysis. It can be applied to study various social phenomena, from information diffusion to disease spread.

Overall, this research provides a valuable tool for accurately estimating personal network size, enabling researchers to gain deeper insights into social dynamics and structures.

Some Apps and Networks We Recommend to Help Stay in Touch and Build Your Network.

Networking is essential for career success, and the digital age has made it easier than ever to connect. A variety of AI-powered tools and apps can help you network more effectively, such as those previously discussed. Such social media platforms like LinkedIn, Instagram, and TikTok can be great places to connect with potential collaborators and clients. There are also some great "old school" networks to be aware of for just about any group of people. Here are some of our favorites in alphabetical order.

American Greetings

American Greetings, a renowned provider of greeting cards and eCards, offers a valuable tool for fostering meaningful connections and strengthening networks. It provides a vast selection of greeting cards for every occasion, allowing you to express your thoughts and feelings in a heartfelt manner. Whether it's a birthday, anniversary, or simply a gesture of appreciation, sending a personalized greeting card demonstrates that you care and remember the people in your life. This can significantly strengthen your personal relationships and create lasting bonds.

These high-quality cards and video cards will elicit a great reaction from your contacts. A helpful tip is to program in the birthdays of your friends and colleagues to receive a reminder of when to send cards to them.

In the digital age, inboxes are constantly bombarded with emails, making it challenging for any message to stand out. However, American Greetings eCards offer a unique and delightful way to break through the clutter and create a memorable experience for recipients.

Blinq

Blinq ditches old-school cardboard for a sleek, customizable digital business card app. Craft your professional identity in minutes, and then share it effortlessly via QR code, link, or even text. Recipients don't need the app, making connections seamless whether you're networking at a conference or exchanging details across continents. The app also works with anyone who has a camera phone to scan the QR code. While more and more apps like LinkedIn, WhatsApp, and WeChat have QR code capabilities, they require more than one click to access. Blinq allows anyone to connect regardless of their preferred apps or type of phone. It doesn't get simpler.

Instagram

Instagram, with its visually-driven platform, has emerged as a powerful tool in the networking toolbox, fostering connections and relationships in unique ways. The platform's emphasis on images and short-form content allows individuals to showcase their personalities, interests, and professional endeavors through a curated visual narrative. In the realm of networking, this serves as a dynamic introduction, providing a snapshot of one's life and professional journey.

The platform's direct messaging feature facilitates direct and immediate communication, enabling individuals to connect effortlessly with like-minded professionals, mentors, or potential collaborators. Whether it's engaging with posts, participating in discussions through comments, or utilizing the direct messaging feature, Instagram offers a versatile space for networking that goes beyond traditional boundaries.

Moreover, the use of hashtags and geotags allows users to discover and connect with individuals who share similar interests or are located in the same geographical area. This fosters niche communities within the platform, creating opportunities for networking with individuals who may not have crossed paths otherwise.

Instagram's Stories feature adds a layer of authenticity to networking efforts. By sharing behind-the-scenes glimpses, updates, and day-to-day experiences, users can build a more personal connection with their audience. This transparency contributes to the development of genuine relationships, a crucial aspect of effective networking.

In essence, Instagram's visual and interactive nature makes it a valuable asset in the networking landscape. It provides a canvas for individuals to express themselves, connect with diverse audiences, and establish meaningful professional relationships that extend beyond the limitations of traditional networking platforms.

Jack and Jill

A beacon of community and empowerment, Jack and Jill of America, Inc. stands as a testament to the vibrancy of Black culture. Founded in 1931, this illustrious social club fosters the growth and development of Black youth aged 8 to 18. Through a diverse range of programs, Jack and Jill cultivates leadership skills, academic excellence, and a strong sense of cultural identity in its members. From character development workshops to scholarship opportunities and community service initiatives, Jack and Jill empowers its young members to become well-rounded individuals and future leaders, leaving a lasting impact on generations of Black youth.

Beyond its focus on youth, Jack and Jill fosters a powerful network of Black families. Parents and guardians actively participate in the club's activities, creating a supportive environment where generations can learn and grow together. This intergenerational exchange strengthens family bonds and fosters a shared commitment to uplifting the Black community. Jack and Jill of America, Inc. is much more than a social club; it's a vibrant tapestry woven from mentorship, cultural pride, and a collective commitment to a brighter future.

LinkedIn

We are very bullish on LinkedIn for unlocking hidden opportunities and strategic networking. In the dynamic and competitive landscape of today's job market, finding the right opportunity requires more than just passively waiting for openings to appear. Proactive networking is an essential strategy for connecting with potential employers, expanding your professional circle, and uncovering hidden opportunities that might not be advertised. LinkedIn, the world's largest professional networking platform, offers a powerful tool for achieving these goals.

However, simply creating a profile and hoping for the best is unlikely to yield significant results. A more strategic approach is required, one that involves leveraging the platform's vast network to your advantage. Reach out to contacts from within organizations with whom you are trying to develop relationships. Look at who those people are connected with and connect with them as well, or at least get a better understanding of who else and which other brands are active in the same space.

In addition, join industry news channels, where you can post comments about relevant topics or amplify others' posts to stand out as an industry thought leader. You can also join alumni clubs within LinkedIn to maintain relationships with old classmates and develop new relationships. Use LinkedIn's recommended profiles and friends of friends to develop even more contacts. Also, subscribe to relevant newsletters on LinkedIn and be sure to look at the robust analytics they supply for every member. It will be revelatory.

The Mighty

The Mighty https://themighty.com/topic/mental-health/ is a website and online community that focuses on sharing personal stories and experiences related to disability and chronic illness. It offers a platform for open and honest discussions about the challenges and triumphs of living with a disability.

The Social Register

The Social Register, a series of regional publications in the United States, has long been associated with exclusivity and wealth. Established in 1886, it aimed to document the supposed "upper class" of American society. Inclusion was based on a combination of factors, including lineage, social prominence, and reputation. Debuts for young women were often announced in the Social Register, signifying their entry into high society.

However, the Social Register's influence has waned considerably in recent decades. Criticisms abound regarding its subjectivity, elitism, and exclusion of people based on race, ethnicity, and religion. Wealth creation through new industries has also shifted the landscape of American society. While some still hold the Social Register in high esteem, it's no longer the sole arbiter of social status in the United States.

TikTok

TikTok, the short-form video platform that has taken the social media world by storm, offers a unique and engaging avenue for networking and building connections. Unlike traditional networking platforms, TikTok's focus on bite-sized, creative content allows users to showcase their personality and expertise in a fun and relatable manner.

The platform's algorithmic nature ensures that content reaches a wide audience, potentially connecting individuals with like-minded professionals or those interested in similar fields. Through creative challenges, duets, and stitches, users can collaborate and engage with each other's content, fostering a sense of community within the TikTok ecosystem.

TikTok's direct messaging feature provides a direct channel for communication, enabling users to reach out to potential collaborators, mentors, or industry peers. The platform's casual and informal environment encourages authentic interactions, breaking down traditional networking barriers.

The use of trending hashtags amplifies the discoverability of content, allowing users to tap into niche communities and connect with individuals who share common interests or professional goals. Whether it's showcasing expertise through educational content or sharing behind-the-scenes glimpses of professional life, TikTok enables users to present a multifaceted view of themselves.

The platform's emphasis on creativity and humor adds an entertaining element to networking efforts. Users can leverage trends and challenges to inject personality into their profiles, making networking a more enjoyable and memorable experience. This lighthearted approach often leads to more meaningful connections as users resonate with authentic and entertaining content.

TikTok's dynamic and entertaining nature transforms networking into a visually compelling experience. It provides a platform for professionals to express themselves creatively, connect with a diverse audience, and build meaningful relationships enjoyably and effectively.

X, Formerly Known As Twitter

X, with its dynamic and real-time nature, serves as a powerful catalyst for fostering connections and expanding professional networks. In the vast landscape of social media, X stands out as a versatile platform that transcends boundaries, allowing individuals to engage with a diverse global audience.

The platform's concise character limit encourages users to communicate succinctly and effectively, making it an ideal space for sharing thoughts, industry insights, and updates. Professionals can leverage this brevity to craft impactful messages that resonate with their audience, sparking conversations and attracting like-minded individuals.

One of X's distinctive features is the ability to connect directly with industry leaders, influencers, and peers. By following and engaging with key figures in your field, you open doors to valuable insights, discussions, and potential collaborations. X's open nature enables you to participate in conversations, share your expertise, and establish yourself as a thought leader in your industry.

Hashtags play a pivotal role on X, serving as virtual meeting points for discussions on specific topics. By strategically using relevant hashtags, individuals can join conversations, discover new connections, and become part of niche communities within their industry. This targeted approach enhances the networking experience, allowing users to connect with professionals who share common interests.

Moreover, X's retweets and like features facilitate the amplification of your content, increasing its visibility and reach. This viral aspect of the platform enables users to organically grow their network as others engage with and share their Xs. This ripple effect can extend your influence far beyond your immediate circle.

X's real-time updates also make it a valuable tool for staying informed about industry trends, news, and events. By actively participating in conversations about these topics, individuals position themselves at the forefront of industry discussions, further enhancing their visibility and networking opportunities.

In essence, X empowers professionals to build and nurture connections on a global scale. Its real-time, concise, and open nature makes it an invaluable tool for networking, enabling individuals to engage with industry leaders, participate in relevant conversations, and establish a strong and influential online presence.

Ethical Considerations and Building Genuine Relationships

Networking is not about exploiting or manipulating others. It's about leveraging powerful networking tools to make meaningful connections with individuals who share your professional interests.

Throughout your interactions, maintain a genuine and respectful approach. Focus on building relationships based on mutual value and shared interests. Avoid being overly pushy or sales-oriented; instead, demonstrate your expertise and willingness to contribute to the organization.

Here is a general overview of how to connect:

- Identify the person with whom you want to connect. Start by narrowing down your target audience. With whom, specifically, do you want to connect? What are their interests, hobbies, and profession? This will help you tailor your approach and find common ground.
- Use online resources. The internet is a powerful tool for connecting with people worldwide. Search social media platforms like LinkedIn, Facebook, Twitter, or Instagram to find individuals who match your criteria. You can also utilize online forums, groups, and communities related to your interests to expand your network.
- Leverage your existing connections. Word-of-mouth and referrals can be highly effective in connecting with people you want to reach. Inform your friends, family, colleagues, and acquaintances about your goal. They might know someone who can introduce you to your target person or provide valuable insights.
- Attend industry events and conferences. Networking events and conferences offer excellent opportunities to connect with like-minded individuals in your field or industry. These gatherings provide a casual setting to approach potential connections, exchange business cards, and engage in meaningful conversations.
- Engage in meaningful conversations. Once you connect with someone, focus on building genuine relationships. Show genuine interest in their experiences, ideas, and perspectives. Active listening,

asking insightful questions, and sharing relevant experiences can foster lasting connections.
- Utilize professional networking platforms. Platforms like LinkedIn and professional networking websites can help you connect with people in your industry or field of interest. Create a professional profile, engage in relevant discussions, and participate in online groups to expand your network.
- Leverage technology to your advantage. Utilize video conferencing tools like Zoom or FaceTime to connect with people virtually, especially if they are located far away. These platforms allow for face-to-face interactions and can strengthen relationships.
- Be patient and persistent. Building meaningful connections takes time and effort. Do not get discouraged if you don't make instant connections. Be persistent in your efforts, and you will gradually build a network of people who can enrich your life and career.

Remember, the key to connecting lies in being genuine, engaging, and persistent. By following these strategies and putting in the effort, you can expand your network worldwide and connect with individuals who can make a positive impact on your life.

AI-Powered Networking Tools

As previously discussed, AI is the elephant in the living room for many businesses and organizations, as well as families. AI is rapidly entering into the networking space. AI-powered networking tools can help you identify potential connections, track your interactions, and even automate some of the networking process. For example, Blinq can help you create a digital business card that you can share with others. When someone scans your Blinq card, they'll be taken to a personalized landing page where they can learn more about you and connect with you on social media or via email.

Artificial intelligence is rapidly transforming all aspects of our lives, and the world of work is no exception. AI is already being used to automate tasks, improve efficiency, and make better decisions. But AI can also play a role in helping people network, collaborate, and deliver on their purpose in life.

One of the biggest challenges of networking is finding the right people with whom to connect. AI can help overcome this challenge by identifying potential connections based on your interests, skills, and experience. For example, LinkedIn uses AI to recommend people you may know, and job boards like Indeed and Glassdoor use AI to match job seekers with open positions.

AI can also help you connect with people in a more meaningful way. For example, AI-powered chatbots can be used to schedule meetings, send follow-up messages, and even have conversations with potential connections on your behalf. This can free up your time to focus on building relationships with the people you meet.

AI can also help people collaborate more effectively. AI-powered tools can be used to translate languages, transcribe meetings, and create shared documents. These features can help teams communicate and work together more efficiently, even if they are located in different parts of the world.

AI can also be used to automate repetitive tasks, freeing up team members to focus on more creative and strategic work. For example, AI-powered tools can be used to generate reports, create presentations, and even write marketing copy.

In today's digital age, building and maintaining a network can feel overwhelming. Thankfully, a wealth of online tools and techniques can help. Professional platforms allow you to showcase your skills and connect with relevant people in your field. Social media platforms, while not solely professional, can be used strategically to connect with like-minded individuals. Online communities built around your interests or industry offer another avenue for meeting new people who share your passions. Additionally, AI-powered tools are emerging that use algorithms to match individuals based on shared skills and goals, fostering potential mentorships, collaborations, or friendships.

Once you've begun building your network, specific tools can help you measure and monitor its growth. Customer relationship management (CRM) software, originally designed for sales teams, can be adapted for personal network management. These tools allow you to track your connections, record interactions, and set reminders for follow-ups. Network visualization tools can help you see your network in a whole new light. By adding connections and categorizing them, you can identify gaps and areas for growth within your network. Social media analytics tools can reveal the impact of your online networking efforts by tracking your reach, engagement metrics, and follower demographics. Finally, relationship-tracking apps can be integrated with your email and calendar to capture data from your interactions with others, helping you build stronger relationships.

Remember, quality matters more than quantity. Focus on building meaningful connections with people who share your values and can contribute to your growth. Authenticity is key—people are drawn to those who are genuine and interested in building real relationships. Don't just take from your network; give back as well. Offer support, share valuable information, and be a resource for others. By leveraging the right digital tools and techniques and by being strategic, authentic, and intentional in your networking efforts, you can build a robust network that empowers you to achieve your personal and professional goals.

Discussion Questions

For Chapter 10, consider these questions:

1. How can you improve or change your current online presence to make broader and generate more contacts?
2. Have you engaged with AI in your daily routine? How? Has it worked well or not?
3. Have you grown your LinkedIn network? If not, what might you do to grow your network?
4. Do you belong to any old-school networks such as Jack and Jill, an alumni organization, or a social club? How do they compare and contrast with your digital communities?

Chapter 11: Old School Matters

Handshakes and coffee dates matter. Handwritten cards and notes matter. There is an enduring value of old-school connecting in a digital age.

The ping of notifications, the endless scroll, the curated feeds—our lives are now soundtracked by the constant hum of the digital world. But amidst the glittering screens and lightning-fast connections, something precious is often lost: the human touch, the warmth of genuine face-to-face interaction, the slow burn of building relationships in the real world. This is where the value of "old-school connecting" shines brightest, a beacon of authenticity in the digital sea.

Beyond the Like Button: Building Deeper Bonds

Sure, social media allows us to connect with a wider circle, but those connections are often fleeting and superficial. A witty comment, a shared meme—they spark momentary engagement but rarely ignite the kind of lasting bond forged through shared laughter over coffee, a heartfelt conversation over a meal, or the unfiltered exchange of ideas in a physical space. Old-school connecting allows us to delve deeper and understand the nuances of body language, the empathy conveyed through a touch, or the unspoken words that hang in the air between two souls.

These are the threads that weave the tapestry of genuine connection, not the pixelated squares on a screen.

In a world where attention is a scarce commodity, traded in bite-sized increments on digital platforms, old-school connecting offers a sanctuary of undivided focus. When we sit across from someone with our phones silenced, our eyes meet,, and we give them the present of our full presence. This focused attention is a rare and valuable currency, nourishing the soul and fostering a sense of trust and respect that cannot be replicated by even the most sophisticated algorithm.

The digital world thrives on ephemeral moments, captured and shared in a flash, and then quickly consumed and forgotten. Old-school connecting, however, creates memories that linger, etched not in the fleeting glow of a screen, but in the shared laughter echoing in a café, the comforting silence of a walk in the park, the spontaneous adventure born from a face-to-face conversation. These shared experiences become touchstones, reminding us of the human connections that enrich our lives and give them meaning.

In the digital age, it is easy to lose ourselves in the virtual labyrinth, neglecting the richness of the real world. So let us rediscover the power of a handwritten note, the joy of an unplanned phone call, and the thrill of a serendipitous encounter in a bustling street. Let us carve out time for coffee dates with friends, board game nights with family, and meaningful conversations with colleagues that go beyond the confines of email threads.

Attending Events, Professional Associations, and Conferences

In the dynamic world of business and professional development, conferences serve as valuable platforms for knowledge exchange, networking, and career advancement. These events provide opportunities to connect with industry experts, potential collaborators, and like-minded individuals. The following is a comprehensive guide to networking effectively at conferences.

Networking at conferences is a strategic endeavor that can significantly contribute to building valuable connections. Before the conference, it is essential to set clear goals for your networking efforts. Identify individuals or companies with whom you wish to connect and conduct thorough research on their backgrounds. This preparation allows you to initiate conversations with relevant and engaging topics.

Arriving early at in-person conferences is crucial for making a positive first impression. Punctuality and a professional appearance set the tone for successful networking. Engaging in active listening and asking insightful questions during sessions demonstrate genuine interest and encourage meaningful conversations.

The exchange of contact information and the importance of a follow-up cannot be overstated. Whether it's collecting business cards at in-person conferences or exchanging numbers, maintaining connections beyond the conference is essential. Sending personalized follow-up notes or emails strengthens the initial connections made during the event.

Participating in social events is a valuable aspect of in-person conferences. Utilize these opportunities to expand your network in a more relaxed setting. Icebreaker activities, conversations, and idea exchanges during these events contribute significantly to relationship-building.

In the context of in-person conferences, additional tips include conducting research on speakers, attendees, and sponsors beforehand. Updating your LinkedIn profile and preparing a concise elevator pitch ensure a strong and memorable introduction. Bringing business cards facilitates the exchange of contact information with new connections.

For conferences, publicizing your attendance on social media platforms before the event using relevant hashtags can help establish connections early on. Reviewing the attendee list, identifying key individuals, and setting specific networking goals contribute to a focused and productive virtual networking experience.

During the conference, active participation in chat and Q&A sessions, attending breakout sessions, and utilizing networking features like one-on-one video chats enhance your visibility and engagement. Having a brief introduction ready, being genuine, and respecting others' time are crucial aspects of successful virtual networking.

The Pen Is Mightier Than the Keyboard

In our hyper-connected world, where emails and texts dominate communication, the art of the handwritten note seems almost quaint. But there's a reason why pen and paper haven't gone completely digital: they offer a unique set of advantages that screens simply can't match.

Unlike laptops and tablets that constantly vie for our attention with notifications and endless browsing, pen and paper provide a distraction-free zone. Studies have shown that this reduced distraction leads to increased focus and deeper engagement with the material at hand. The act of writing itself also strengthens the connection between the brain and the information being written, improving memory and recall when you revisit your notes later.

A blank page is an invitation for exploration. Unlike the limitations of a digital format, pen and paper allow your thoughts to flow freely. You can create mind maps, visual organizers, and doodles to capture ideas and spark connections that might be stifled by a sterile digital environment.

In a world saturated with digital communication, a handwritten note stands out as a thoughtful gesture. It conveys a level of care and personalization that a typed message simply can't replicate.

The Allure of Customized Stationery

Beyond the cognitive benefits, there's something undeniably charming about customized stationery. A notebook adorned with your favorite quote or a folder with an inspiring design can make the note-taking process more enjoyable. Even a beautiful pen can elevate the act of writing from a chore to a ritual.

The key is not to see pen and paper as rivals to technology, but rather as complementary tools. Tablets with styluses, for example, offer the focus benefits of handwriting with the searchability and organization of digital notes.

Ultimately, the choice between digital and handwritten depends on your learning style and preferences. But in an age of information overload, there's something to be said for the focus, creativity, and personal touch that pen and paper can bring. So why not pick up a pen and rediscover the power of the written word?

Remember, it's not about abandoning the digital world entirely but finding a healthy balance. Use the power of the internet to reach out and connect, and then nurture those connections through the warmth of personal interaction. So, go ahead, pick up the phone, schedule that coffee date, and watch your career flourish under the sunshine of genuine connection.

Advice from David Rockefeller

David Rockefeller, a prominent philanthropist and the grandson of oil baron John D. Rockefeller Sr., was known for his vast network of contacts. Toby even had a few occasions to join him and his family for lunches and dinners given his friendship of many years with Susan and David Rockefeller. In addition to his charm, grace, and intellectual curiosity, he was also very responsible with his personal and professional networks. He meticulously recorded information about these contacts on 3 by 5" index cards, which he stored in a custom-built five-foot high Rolodex machine. This Rolodex contained contact information for about 200,000 people around the world, including heads of state, business titans, and celebrities.

Mr. Rockefeller's Rolodex offered a unique glimpse into his life and work. It showed the depth of his connections with world leaders like Anwar Sadat and Leonid Brezhnev and his personal relationships with figures like Henry Kissinger and John F. Kennedy. The Rolodex cards also documented Rockefeller's long career at Chase Manhattan Bank, where he helped forge business relationships in countries like China and Iran.

Beyond just names and numbers, Mr. Rockefeller's Rolodex was a testament to his dedication to relationship building. He reportedly recorded details of every meeting he had on the index cards, allowing him to refresh his memory before encounters even years later. This meticulous record-keeping impressed his colleagues. James Wolfensohn, a friend and former World Bank president, said Rockefeller could pick up a conversation as if it happened the day before, thanks to his "extraordinary record system."

Mr. Rockefeller's Rolodex wasn't static either. He updated it constantly, reflecting the ever-evolving nature of his relationships. For instance, one of Henry Kissinger's cards mentioned when he was knighted and included a note emphasizing that he shouldn't be addressed as Sir Henry. This anecdote highlights Mr. Rockefeller's attentiveness to detail and his desire to maintain a close connection with his contacts.

A 2017 *Wall Street Journal* article by Joanne S. Lublin noted: "Some might say David Rockefeller, a scion of America's greatest fortune and the veteran chief executive of Chase

Manhattan Bank, was a dedicated networker long before the age of Facebook."

Lublin also cites Nancy Koehn, a Harvard business professor and historian, who stated: "In the annals of CEO history, the breadth and depth of this record of contacts stand out. This is a man with a large, long reach."

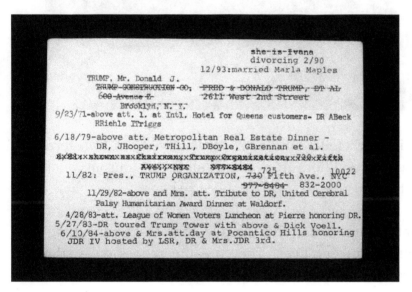

David Rockefeller kept a custom-built Rolodex machine that stored about 200,000 3-by-5-inch cards with contact information and a record of every meeting he had with about 100,000 people from around the world, including this one on Donald Trump.

David Rockefeller's Rolodex cards have been transferred to file drawers. Photo from The Wall Street Journal, Benjamin Hoste.

Relationships form the threads that weave meaning and resilience in our lives. They are the echoes of laughter in shared moments, the unwavering support in times of need, and the whispers of encouragement that propel us forward. Yet, in our fast-paced world, nurturing these precious threads can often fall by the wayside. As we have reviewed in this book, there is an etiquette to networking.

Cleared for Takeoff: Building Strong Relationships in a Chaotic World

Pre-Flight Check: Authenticity is Key

Our connections soar highest not in phony greetings and fleeting interactions, but in the fertile ground of being genuine. Strong relationships are built on mutual respect, a thirst for knowing the real person, and the courage to be vulnerable. They're safe spaces where we can ditch the masks, share our triumphs and woes, and be accepted for who we truly are, even for the turbulence we experience. In contrast, inauthentic connections are transactional, often fueled by convenience or what benefits one gets. They lack depth, clear communication, and the willingness to invest time and effort. While they might offer temporary company, they fail to provide the support and growth that true connection offers.

Maintaining Altitude: Staying Connected

Keeping relationships strong requires effort, but the rewards are immeasurable. Here are some tips to keep those wings of connection well-maintained:

- Embrace the power of presence: In a world dominated by digital chatter, prioritize face-to-face interactions. Share a meal, explore together, or simply have a meaningful conversation.
- Become a master communicator: Go beyond small talk. Ask insightful questions, truly listen, and offer genuine

support. Remember, sometimes the most powerful connections happen in comfortable silence.
- Celebrate and commiserate: Share in your loved ones' victories, and offer a shoulder to cry on during their struggles. Being there in sunshine and rain strengthens the fabric of your connection.
- Small gestures, big impact: A handwritten note, a surprise call, or a thoughtful gift can speak volumes. These gestures show you care and keep the connection alive.
- Technology for good: While virtual interactions shouldn't replace real-life connections, technology can be a valuable tool to stay in touch. Schedule regular video calls, send encouraging messages, or share meaningful content.

Nurturing the Journey: Deepening Bonds

Connections aren't static; they require constant care and attention. Here are some ways to keep those bonds growing:

- Invest in shared experiences: Create memories together. Take a trip, pursue a shared hobby, or simply explore your local area. Shared experiences strengthen bonds and provide fodder for future conversations.
- Practice forgiveness: No flight is perfect. There will be mistakes and misunderstandings. Learn to forgive and move on, allowing your connection to grow stronger through vulnerability and understanding.

- Express gratitude: Let your loved ones know how much they mean to you. Appreciation strengthens bonds and fosters a sense of mutual respect and value.
- Make time for individual connections: While group gatherings are important, prioritize one-on-one interactions. Dedicate time to each person, allowing you to deepen your understanding and strengthen individual bonds.

The Ripple Effect: A Network of Well-Being

Strong relationships aren't just personal luxuries; they're the foundation of well-being. They provide us with a sense of belonging, security, and purpose. Studies show that strong social connections can boost our physical and mental health, reduce stress, and improve overall life satisfaction. When we invest in our relationships, we not only enrich our own lives but also contribute to a more positive and connected world.

So let's tend to the network of our relationships with care and intention. Let's celebrate the joy of connection, offer support in times of need, and nurture the threads that bind us together. For in the intricate network of belonging, we find meaning, support, and the strength to face life's challenges and build a future brighter than any single flight could ever be.

Here are some specific examples of how the famous connectors listed previously maintain their relationships:

- Emma Chamberlain stays in touch with her friends and family by texting, video chatting, and scheduling regular visits. She also makes sure to post about them on social media and let them know how much she appreciates them.
- Billie Eilish has built a strong community of fans through her music and social media. She interacts with her fans regularly by responding to comments and messages, and she even hosts live Q&A sessions.
- Malala Yousafzai has used her platform to connect with other activists and advocate for girls' education. She has also traveled all over the world to meet with young people and encourage them to stand up for their rights.
- Amanda Gorman has used her poetry to connect with people from all walks of life. She has performed at events for a variety of organizations, and she has also met with world leaders to discuss issues such as social justice and climate change.
- Warren Buffett is known for his humility and commitment to giving back to his community. He has donated billions of dollars to charity and has encouraged other wealthy people to do the same.
- Oprah Winfrey is known for her generosity and ability to connect with people on a personal level. She has used her platform to help others and promote positive change in the world.

By following these tips and learning from these famous connectors, you can build and maintain strong relationships in all aspects of your life. You can experiment with finding the right balance between digital and analog networking. You can even shift the balance over time. The important thing to remember is just get started—put a stake in the ground and move forward from there, one relationship at a time.

If you want to maintain strong relationships, it's important to be authentic, consistent, and regular. Be yourself, be reliable, and make time for the people you care about. Communicate openly and honestly, be supportive and understanding, and show your appreciation. By following these tips, you can build and maintain strong relationships that will last a lifetime.

A Word About Letting Go

Another aspect of the natural evolution of relationships is sometimes the act of letting go. Embracing change sometimes means letting go of some relationships which, like life itself, are ever-evolving entities. Relationships undergo constant transformation, shaped by the passage of time, the accumulation of experiences, and the shifting dynamics of individuals. While some connections deepen and strengthen over the years, others may gradually fade, their intensity diminishing until they eventually dissolve into a memory. This natural waning of relationships, often perceived as a loss or a failure, can instead be viewed as an inevitable and necessary part of the human experience.

Accepting that relationships can and do sometimes wane is not a resignation to apathy or indifference. It is, rather, an acknowledgment of the impermanence of things, a recognition that change is an inherent aspect of life. Just as the seasons shift, and the leaves on the trees change color and fall, relationships, too, undergo transformations. Some connections may blossom into lifelong partnerships while others may serve their purpose for a specific time and then fade away.

The key to navigating these transitions lies in the ability to let go gracefully, recognize when a relationship has run its course, and allow it to dissolve without clinging to what was once familiar and comfortable. Dwelling on past glories or holding onto expectations that can no longer be met only prolongs an inevitable end and hinders personal growth.

Letting go of relationships that have waned is not an act of coldness or abandonment; it is an act of self-preservation and emotional maturity. It is about recognizing that both individuals have evolved, their paths may have diverged, and their needs may have changed. Holding on to a connection that no longer serves either person is akin to clinging to a life raft in a storm that has long passed.

While letting go can be challenging, it also opens the door to new possibilities and opportunities for growth. It allows individuals to pursue new connections, explore different interests, and rediscover themselves. It is a chance to embrace the ever-changing landscape of life and move forward with a renewed sense of purpose and optimism.

The ability to let go and move forward is not limited to romantic relationships; it extends to all forms of human interaction. Friendships, family ties, and even professional connections can undergo similar transformations. Embracing change and accepting that relationships evolve over time is essential for maintaining healthy and fulfilling bonds.
In any relationship, it is crucial to strike a balance between cherishing the past and embracing the future. While reminiscing about shared memories can be a source of joy and connection, dwelling too much on the past can hinder the relationship's natural progression. It can prevent individuals

from adapting to changing circumstances and fostering new experiences.

Moving forward, or "advancing the narrative" as Toby often advises, requires a willingness to embrace the present and actively shape the future of the relationship. It involves open communication, a shared understanding of expectations, and a commitment to mutual growth. It means nurturing the connection, not out of nostalgia or obligation, but out of a genuine desire to maintain a healthy and fulfilling bond.

By recognizing when a connection has run its course and gracefully letting go, we open ourselves to new possibilities, personal growth, and the chance to forge deeper, more meaningful relationships in the future.

Remember, networking is about building mutually beneficial relationships, not just collecting contacts. By following these etiquette guidelines, you can create meaningful connections that can lead to personal and professional success. Networking in the digital age is easier than ever. By following our suggested tips, you can use social media, online events, and virtual networking to connect with people from all over the world and achieve your professional and personal goals.

Discussion Questions:

For Chapter 11, consider these questions:

- What are some tips for maintaining relationships?
- Have you ever lost a relationship or had it worsened only to realize too late that you had neglected it? What did you learn from that experience that you bear in mind with current relationships?
- In an era of polarization, how have you managed to maintain relationships with others who have diametrically opposite views from yours?

Building Community

Image used under license from Shutterstock.com

Chapter 12: The Power of Connection: Optimizing Your Network and Legacy

Connection is a fundamental human need. We are wired to connect with others, and our relationships play a vital role in our overall well-being. Studies have shown that people who have strong social connections are happier, healthier, and live longer.

The power of networking extends far beyond the realm of professional benefits and social circles. As pilots, we have learned that each stop along the way not only accumulates hours in our logbooks but also enriches our lives with the communities and activities we experience every time. Our logbooks are a sort of tapestry for any pilot.

It weaves a deeper tapestry, enriching our lives with purpose, meaning, and a lasting legacy. Let's unravel these threads and discover how connection transcends mere career advancement, touching the very core of our human experience.

Your Tapestry or Logbook of Purpose

Through diverse connections, we encounter a kaleidoscope of perspectives, passions, and challenges. This exposure broadens our horizons, sparking a deeper understanding of ourselves and the world around us. As we engage with these different threads, we begin to identify causes that resonate with our values, sparking a sense of purpose that transcends personal ambition. Whether it's advocating for environmental protection, mentoring young minds, or building bridges between communities, our connections guide us toward fulfilling contributions that leave a lasting mark.

As you have noted throughout this book and your own lived experience, life as a tapestry is not an individual masterpiece; it's a collaborative art form. By weaving ourselves into the fabric of connection with others we meet along the way, we become part of something larger than ourselves. We find meaning in shared experiences, triumphs, and struggles. Celebrating a colleague's promotion, offering comfort during a friend's loss, or collaborating on a community project—these moments infuse our lives with a sense of belonging and shared purpose. They remind us that our actions, however small, contribute to a collective tapestry of meaning that transcends individual achievements.

The legacy we leave behind is not etched in monuments or inscribed on plaques. It's woven into the lives of those we touch, the communities we strengthen, and the impact we make on the world. Through our connections, we become mentors, role models, and pillars of support, leaving behind a ripple effect of positive change. Whether it's inspiring a younger generation through our experiences, advocating for a cause we believe in, or simply leaving a trail of kindness in our wake, our connections ensure that our legacy lives on even after the final thread is spun.

As a pilot logs hours into their logbook through the years, so do we add to our network, impact, and legacy. We cultivate purpose, meaning, and legacy with every daily interaction with others.

We must actively nurture our connections:

- Embrace authentic interactions: Connect with people on a deeper level, beyond professional titles and social facades. Share your passions, values, and dreams. These genuine connections form the strongest threads in the tapestry.
- Seek opportunities for mutual growth: Encourage others, offer support, and celebrate their achievements. Be a catalyst for positive change in their lives, knowing that their success enriches the entire web.
- Contribute to causes you care about: Join forces with others to address challenges and create a better future. Leave your mark on the world through collective action, knowing that your contribution, however small, strengthens the tapestry of meaning.
- Live with intention and mindfulness: Every thread you weave, every interaction you foster, contributes to the tapestry of your life. Be conscious of the impact you make, strive for kindness, and embrace the interconnectedness of all things.

Remember, the tapestry of life is not woven in isolation. Like our flight across America, it's a collaborative masterpiece, a vibrant dance of connections that imbues our lives with purpose, meaning, and a legacy that reaches far beyond the individual. This book is a creative result of a connection seeded 17 years ago. So, let us embrace the power of connection, actively weave the threads of belonging, and leave behind a world richer, more vibrant, and filled with the echoes of our shared humanity.

This expanded version emphasizes the powerful link between networking and a fulfilling life, highlighting how fostering connections brings purpose, meaning, and a lasting legacy. By actively nurturing our relationships and contributing to something larger than ourselves, we become architects of a life woven with significance and leave a positive mark on the world. We hope this resonates with you and further enhances your understanding of the transformative power of responsible networking.

Master Planning: Strategies for Effective Networking and Building Your Legacy

To maximize the benefits of networking, companies should adopt a strategic approach that includes:

- Defining goals: Clearly defined networking goals provide direction and ensure that networking efforts are aligned with overall business objectives.
- Identifying a target audience: Understanding the ideal customer personas, partners, or investors helps companies focus their networking efforts on the most relevant individuals and groups.
- Actively participating: Engaging in networking events, joining online communities, and actively participating in discussions demonstrate a company's commitment to building relationships and expanding its network.

- Nurturing relationships: Networking is a continuous process, requiring consistent effort to maintain and strengthen connections. Regular communication, follow-ups, and expressing appreciation go a long way in building lasting relationships.

Networking is a powerful tool that can propel businesses toward success. By embracing strategic networking practices, companies can cultivate meaningful connections, expand their reach, and achieve their growth objectives. As the business landscape continues to evolve, the importance of networking will only grow, making it an essential element for companies seeking to thrive in the ever-changing world of commerce.

AI can also help people deliver on their purpose in life. For example, AI-powered tools can be used to identify your strengths and weaknesses, develop your skills, and set goals. AI can also help you find opportunities that align with your values and interests.

Networking is a two-way street. Be prepared to offer value to your connections as well. Share relevant industry news, provide helpful insights, or offer assistance with projects or tasks. By demonstrating your willingness to contribute, you'll deepen your relationships and establish yourself as a valuable asset.

Throughout this book, we have explored the many benefits of intentional networking, from career advancement to personal enrichment. We have seen how networks can help us learn new things, find new opportunities, and build meaningful relationships.

The Power of Connection

When we connect with others, we are not just exchanging information or making business deals. We are building relationships. We are creating a community. It is this community that gives us the strength, support, and inspiration to achieve our goals.

Think about your own life. Who are the people who have helped you along the way? Who has given you advice, encouragement, and support? Who has challenged you to grow and be better?

Chances are, these people are part of your network. Your network is your support system. It is your source of strength and inspiration. It is the community that helps you achieve your goals. It is the power of connection that makes it all possible.

Career Growth

In the book *The Formula: The Universal Laws of Success*, Albert-László Barabási emphasizes the importance of networking for career growth. He argues that our networks are our most valuable assets and that we should invest time and effort in building and maintaining them.

Barabási provides a number of insights into how networking can help us succeed in our careers:

- Networks give us access to information and opportunities. When we connect to a diverse group of people, we learn about new things and have access to opportunities that we would not otherwise know about.
- Networks help us build our reputations. When we interact with people in our networks and make positive contributions, we build our reputations as knowledgeable, reliable, and helpful professionals.
- Networks can help us get ahead at work. When we have strong relationships with our colleagues, managers, and other stakeholders, we are more likely to be considered for promotions, raises, and other opportunities.

Barabási also provides some tips for effective networking:

- Be genuine and authentic. People can tell when you are being fake, so it is important to be yourself and build genuine relationships.
- Be helpful and supportive. Offer to help others in your network, and be there for them when they need you.
- Be active and engaged. Attend industry events, participate in online forums, and regularly reach out to people in your network.

Here are some specific examples of how networking can help with career growth, as mentioned in *The Formula*:

- Getting a new job: When you are looking for a new job, your network can be a valuable resource. You can ask your contacts for referrals, learn about job openings that are not publicly advertised, and get advice on how to interview and negotiate.
- Getting promoted: When you are up for a promotion, your network can help you advocate for yourself and make your case to your manager. Your contacts can also provide you with feedback on your performance and help you identify areas for improvement.
- Starting a business: If you are thinking about starting your own business, your network can be a valuable source of support and advice. You can tap into your contacts for expertise in different areas, such as finance, marketing, and sales. You can also use your network to raise capital and find customers.

Overall, *The Formula* makes a compelling case for the importance of networking for career and personal growth. By building and maintaining strong relationships with others, we can access information and opportunities, build our reputations, and get ahead in our careers.

Pre-flight Checklist

How does one maintain and nurture relationships effectively? Do you ever think about all of the relationships that you have passed on over the years because you didn't have the will or skill to keep them. Have you asked yourself if you might have remained in vibrant relationships if you were better at nurturing them?.

From a ten-thousand-foot level, networking for us is about putting yourself in a position to meet people, developing a process to stay engaged, auditing your network periodically and leaning into the relationships that you care to expand. Like exercising our bodies at the gym, vibrant relationships require regular exercise. There are plenty of digital and analogue ways to facilitate this, as we have discussed throughout this book.

Our suggestion for your checklist to keep handy, just as every pilot keeps a pre-flight checklist in the cockpit to use before every flight, might include the following:

Mindset and Communication:
- Focus on Giving: Build relationships by offering help and support, not just seeking favors.
- Be Genuinely Interested: Ask questions, listen actively, and show you care about the other person's life.
- Practice Empathy: See things from their perspective and acknowledge their feelings.

- Delight: Find fun, creative ways to acknowledge someone in a way that cuts through the clutter we all find in our in-boxes. Perhaps sharing a photo from a past shared experience, or a funny cartoon that reminds you of a shared moment.
- Communicate Clearly: Express yourself well and be an attentive listener.

Practice Connecting:
- Become a Connector: Introduce people in your network who might benefit from knowing each other.
- Schedule Regular Catch-Ups: Maintain contact with people through calls, emails, or in-person meetings.
- Expand Your Connection: Over time, get to know the friends and family that your contact has discussed in the past. Give them some space in your relationship.
- Celebrate Milestones: Acknowledge and celebrate achievements and important events in your contacts' lives.
- Offer Help Proactively: Look for opportunities to assist others without being prompted.
- Be Reliable and Trustworthy: Follow through on commitments and be someone people can depend on.
- Express Gratitude: Thank people for their time, support, and contributions.

Technology and Social Media:
- Utilize Technology: Leverage online tools like social media to stay connected and share updates.

- Personalize Communication: Don't rely solely on generic messages, personalize your interactions.
- Maintain a Positive Online Presence: Be mindful of what you post and how it might affect your relationships.
- Use Old School Methods: Still send a postcard or thank you note via the postal service periodically. People appreciate the gesture that goes beyond a digital message.

Balance and Quality Time:
- Focus on Quality over Quantity: Prioritize deep, meaningful interactions over superficial connections. But also know who merits more time and attention in your network. Not everyone has the same importance to you and it is appropriate for you to believe this and apply it in your networking.
- Maintain Healthy Boundaries: Don't overload yourself or others with requests, respect limitations.
- Enjoy the Process: Building relationships should be enriching, focus on the positive aspects of connection.

A Word About Polarization

Polarization is a growing problem in our society, and it can be difficult to see how to bridge the divide. However, one way to start is by connecting the dots through professional and personal networking.

When we network, we have the opportunity to meet people from all walks of life. We learn about their experiences, their perspectives, and their values. This can help us better understand the world around us and appreciate the diversity of human experience.

In addition, networking can help us build relationships with people who have different viewpoints. When we have relationships with people who disagree with us, we are more likely to be open to their ideas and listen to their concerns. This can lead to more productive conversations and a better understanding of the issues that divide us.

Networking can help us break down stereotypes. When we meet people from different backgrounds, we realize that they are not just one-dimensional stereotypes. We see them as individuals with their own unique experiences and perspectives. This can help us overcome the prejudices that can lead to polarization.

Networking can help us find common ground. Even though we may disagree with people on some issues, we are likely to have some things in common. Networking can help us identify those shared interests and build relationships based on them. This can create a foundation for trust and respect, even when we disagree.

Networking can help us learn from each other. When we talk to people who have different viewpoints, we learn about their experiences and perspectives. This can help us better understand the issues that divide us and see things from a different point of view. This can lead to more informed and nuanced opinions.

Networking can help us build bridges between communities. When we build relationships with people from different backgrounds, we help connect our communities. This can lead to more collaboration and cooperation, which can help solve common problems.

Going from the Particular to the Universal

The ability to go from the particular to the universal is a fundamental skill that is essential for learning, problem-solving, and creativity. It allows us to see the patterns and connections that underlie individual experiences and apply our knowledge to new situations.

Scientists use induction and abduction to develop theories and hypotheses about the natural world. For example, Charles Darwin observed a variety of different plants and animals around the world, and he used his observations to develop the theory of evolution.

Artists use their personal experiences and observations to create art that resonates with a universal audience. For example, the paintings of Vincent van Gogh often depict the simple life of peasants, but they also convey a sense of universal human emotion.

Authors use their writing to explore universal themes such as love, loss, and redemption. For example, the novel *To Kill a Mockingbird* by Harper Lee tells the story of a young girl living in the American South during the Great Depression, but it also explores universal themes such as racism and injustice.

As you venture out into the world, try to see the patterns and connect the dots, going from the particular to the universal.

Pay attention to patterns and connections. When you observe the world around you, look for patterns and connections between different things. For example, if you notice that a lot of successful people are hard workers, you might infer that hard work is the key to success.

Ask yourself why. When you observe something, ask yourself why it is happening. For example, if you see a homeless person on the street, you might ask yourself why they are homeless. This can help you develop a deeper understanding of the world around you and identify the root causes of problems.

Be open to new ideas. Don't be afraid to challenge your existing beliefs and consider new perspectives. This will help you see the world in new ways and come up with new, innovative solutions to problems.

Going from the particular to the universal is a powerful skill that can help us learn, grow, and make a difference in the world. By paying attention to patterns and connections, asking ourselves why, and being open to new ideas, we can develop a deeper understanding of the world around us and create solutions that benefit everyone.

As this book has demonstrated, connecting the dots through networking, going from your particular to the universal because you have your higher purpose as your North Star, you will optimize your life and legacy.

You will learn from the experiences of others. When we connect with people from different backgrounds and with different perspectives, we can learn from their experiences and broaden our own understanding of the world. This can help us see the world in new ways and come up with new and innovative solutions to problems.

You will identify patterns and connections. When we network with people from different fields and industries, we can start to see patterns and connections that we might not have noticed on our own. This can help us develop a deeper understanding of the underlying forces that shape our world.

You will gain access to new resources and opportunities. When we network with people who share our goals and interests, we can gain access to new resources and opportunities that we might not have had on our own. This can help us achieve our goals more quickly and effectively.

For example, a scientist who is working on a new drug might network with other scientists in the field to learn from their experiences and gain access to new resources. An artist might network with other artists and art galleries to get their work seen by a wider audience. An entrepreneur might network with other entrepreneurs and investors to raise capital for their new business. A social worker who is working with a homeless client might network with other social workers and homeless service providers to learn about best practices and identify resources that can help their client. A teacher who is developing a new lesson plan might network with other teachers and educational experts to get feedback on their plan and identify classroom resources.

By defining your sense of purpose in life and pursuing it by building relationships with people from different backgrounds and with different perspectives, you will learn from their experiences, identify patterns and connections, and gain access to new resources and opportunities. This will help you achieve your goals more quickly and effectively and make a positive impact on the world.

Why It Matters

Addressing universal challenges and opportunities and collaborating with like-minded individuals are fundamental human skills that allow us to learn, grow, and evolve. These abilities are also essential for the evolution of civilizations. When we are able to explore and then define our higher purpose and see the patterns and connections that relate to it, we can develop a deeper understanding of ourselves and the world around us. We can also use this understanding to create new ideas and solutions that benefit everyone.

We evolve as human beings by learning from our experiences and adapting to new challenges. International networking allows us to do this more effectively. For example, if we experience a personal setback, we can learn from it and become more resilient. We can also use our experience to help others who face similar challenges. It also allows us to develop empathy and compassion for others. When we can see the world from another person's perspective, we are more likely to understand and care about their experiences. This empathy and compassion are essential for building strong relationships and creating a more just and equitable world.

Civilizations evolve through the accumulation of knowledge and the development of new technologies. But human beings are always the common denominator of such evolution.

Tapping into Our Shared Humanity

When we network with intention, going from the particular to the universal, we tap into our shared humanity. We realize that despite our individual differences, we all share the same basic needs and desires. We all want to be loved and respected. We all want to live in a safe and peaceful world.

This realization of our shared humanity is essential for building a better world for everyone. When we recognize that we are all interconnected, we are more likely to work together to solve common problems. We are also more likely to treat each other with respect and compassion.

We embarked on this journey in *Connecting the Dots* with a series of fundamental questions: Who are we? Who are the people we weave into the fabric of our lives? What path do we choose to walk? What legacy will we leave behind?

Often, these inquiries simmer beneath the surface, answered by happenstance rather than intention. Here, we've explored a proactive approach, equipping you with methods and real-life examples to answer these crucial questions with honesty, efficiency, and, most importantly, a sense of fulfillment.

But this exploration wasn't a solitary one. We delved into the intricate web of your human connections, both existing and potential. By deliberately exploring your network, you've uncovered patterns, navigated pitfalls, and seized opportunities you may have never noticed before.

You've seen how, in today's dynamic world, building and nurturing relationships isn't just a virtue; it's a cornerstone of success and happiness. In a landscape fueled by collaboration and connection, the ability to effectively navigate interpersonal interactions is paramount to personal and professional growth.

Strong networks function as conduits to a wealth of knowledge, resources, and opportunities. They open doors to fresh ideas, collaborative ventures, and partnerships that spark innovation and propel growth. Yet, in a digital age that can sometimes isolate us, meaningful human connection fosters a sense of belonging, support, and community — essential elements for our well-being and ability to weather life's storms.

We've learned that networking and interpersonal engagement are not solely about building social capital; they are conduits for cultivating empathy, understanding, and respect for others. By connecting with people from diverse backgrounds and perspectives, we broaden our worldviews, challenge our assumptions, and become adept at navigating a world increasingly interwoven.

The ability to connect with others is more valuable than ever in this era of uncertainty. By consciously investing in our networks and interpersonal skills, we invest in our own future, our careers, our relationships, and the communities to which we belong. In essence, we are laying the groundwork for a world that's more connected, collaborative, and ultimately, more successful. After all, "connecting the dots" isn't just a trendy phrase; it's a core concept that fuels human understanding, empowers problem-solving, and guides us in our search for significance.

Now, as you step forward from these pages, remember that the web of connection is ever-evolving. Continue to nurture your existing relationships, reach out to new people, and embrace the growth and opportunities that these connections bring. May the logbook you fill each day be full of rich and vibrant flights, each one a testament to the power of human connection. And may the hours in flight—your time on this planet—be purpose-driven and always take you closer to the legacy you alone can achieve. With your legacy as your destination, your journey will be joyful, impactful, and profound. Excelsior!

Discussion Questions

For Chapter 12, consider these questions:

- Do you have a sense of purpose or know someone who does? How did it come about?
- Have you tried to ask ChatGPT or Gemini.Google.com to help you define your purpose?
- Who are deceased people, famous or not, that you admire for their legacy? Why?

Acknowledgments

Barron Thomas, Harlan Bratcher, Pooja Kanuga, Barron Thomas, Joan Ai, Jaha Cummings, Julian Leone, Edward Bergman.

About the Authors

- Toby Usnik is a communications professional with thirty years of expertise in strategic communications. He has held leadership roles in multinational organizations that include the U.K. government, Christie's, The New York Times and American Express. A frequent public speaker, he is author of "The Caring Economy: How to Win With CSR" and hosts a weekly podcast by the same name on all major platforms, as well as YouTube, LinkedIn and TikTok. He is also a sought-after career coach and instructor on AI and Large Language Models for social impact. In his free time, he enjoys flying and spending time with his husband and two rescue Jack Russell Terriers.
- Samir Kanuga has earned a reputation as a seasoned CFO with expertise in driving strategic growth for global organizations, emphasizing quality performance, and customer success. Easily able to move from high-level planning and visioning to face-to-face stakeholder engagement and staff training, he ensures that companies and shareholders enjoy the best possible results. He is very close to his parents, Sunil and Rita. His dad was instrumental in helping him pursue his start in aviation, and Samir worked with his dad for well over a decade in paper manufacturing. In his free time, he enjoys flying, teaching flying, riding his Peloton, and spending time with his family (his wife Pooja and three kids). Samir has a lifelong enthusiasm for aviation and is a FAA-

rated Airline Transport Pilot & Certificated Flight Instructor.

# # #

Printed in the USA
CPSIA information can be obtained
at www.ICGtesting.com
CBHW081147070624
9723CB00044B/784